Applied Mathematical Sciences

Volume 205

Series Editors

Anthony Bloch, Department of Mathematics, University of Michigan, Ann Arbor, MI, USA
abloch@umich.edu
C. L. Epstein, Department of Mathematics, University of Pennsylvania, Philadelphia, PA, USA
cle@math.upenn.edu
Alain Goriely, Department of Mathematics, University of Oxford, Oxford, UK
goriely@maths.ox.ac.uk
Leslie Greengard, New York University, New York, NY, USA
Greengard@cims.nyu.edu

Advisory Editors

J. Bell, Center for Computational Sciences and Engineering, Lawrence Berkeley National Laboratory, Berkeley, CA, USA
P. Constantin, Department of Mathematics, Princeton University, Princeton, NJ, USA
R. Durrett, Department of Mathematics, Duke University, Durham, CA, USA
R. Kohn, Courant Institute of Mathematical Sciences, New York University, New York, NY, USA
R. Pego, Department of Mathematical Sciences, Carnegie Mellon University, Pittsburgh, PA, USA
L. Ryzhik, Department of Mathematics, Stanford University, Stanford, CA, USA
A. Singer, Department of Mathematics, Princeton University, Princeton, NJ, USA
A. Stevens, Department of Applied Mathematics, University of Münster, Münster, Germany
S. Wright, Computer Sciences Department, University of Wisconsin, Madison, WI, USA

Founding Editors

F. John, New York University, New York, NY, USA
J.P. LaSalle, Brown University, Providence, RI, USA
L. Sirovich, Brown University, Providence, RI, USA

The mathematization of all sciences, the fading of traditional scientific boundaries, the impact of computer technology, the growing importance of computer modeling and the necessity of scientific planning all create the need both in education and research for books that are introductory to and abreast of these developments. The purpose of this series is to provide such books, suitable for the user of mathematics, the mathematician interested in applications, and the student scientist. In particular, this series will provide an outlet for topics of immediate interest because of the novelty of its treatment of an application or of mathematics being applied or lying close to applications. These books should be accessible to readers versed in mathematics or science and engineering, and will feature a lively tutorial style, a focus on topics of current interest, and present clear exposition of broad appeal. A compliment to the Applied Mathematical Sciences series is the Texts in Applied Mathematics series, which publishes textbooks suitable for advanced undergraduate and beginning graduate courses.

More information about this series at http://www.springer.com/series/34

Asen L. Dontchev

Lectures on Variational Analysis

Asen L. Dontchev
University of Michigan–Ann Arbor
Ann Arbor
MI, USA

ISSN 0066-5452 ISSN 2196-968X (electronic)
Applied Mathematical Sciences
ISBN 978-3-030-79910-6 ISBN 978-3-030-79911-3 (eBook)
https://doi.org/10.1007/978-3-030-79911-3

Mathematics Subject Classification: 49-02, 90-02, 49J52, 49J53, 49K40, 90C31

© The Editor(s) (if applicable) and The Author(s), under exclusive license to Springer Nature Switzerland AG 2021
This work is subject to copyright. All rights are solely and exclusively licensed by the Publisher, whether the whole or part of the material is concerned, specifically the rights of translation, reprinting, reuse of illustrations, recitation, broadcasting, reproduction on microfilms or in any other physical way, and transmission or information storage and retrieval, electronic adaptation, computer software, or by similar or dissimilar methodology now known or hereafter developed.
The use of general descriptive names, registered names, trademarks, service marks, etc. in this publication does not imply, even in the absence of a specific statement, that such names are exempt from the relevant protective laws and regulations and therefore free for general use.
The publisher, the authors, and the editors are safe to assume that the advice and information in this book are believed to be true and accurate at the date of publication. Neither the publisher nor the authors or the editors give a warranty, expressed or implied, with respect to the material contained herein or for any errors or omissions that may have been made. The publisher remains neutral with regard to jurisdictional claims in published maps and institutional affiliations.

This Springer imprint is published by the registered company Springer Nature Switzerland AG
The registered company address is: Gewerbestrasse 11, 6330 Cham, Switzerland

To Terry Rockafellar,
mathematician, mentor, and friend

Preface

Variational analysis can be briefly described as the mathematics of optimization and control. The name became commonly accepted after the publication of the book *Variational Analysis* by R. T. Rockafellar and R. J-B. Wets in 1998. However, basic concepts, ideas, and results that are now regarded as being in the heart of the area have been around much longer. Variational analysis unifies theories and techniques that have been developed in *calculus of variations*, *mathematical programming*, and *optimal control* and covers areas such as *convex analysis, nonlinear analysis, nonsmooth analysis,* and *set-valued analysis*. In a separate direction, it has been defined to include topics in differential geometry and functional analysis. It is a good example of mathematics with deep roots, rich theory, and a variety of applications, whose rapid growth in the last decades has been driven not only by the curiosity of researchers but also by very practical problems such as financial planning or steering a flying object.

The aim of this book is to present an introduction to variational analysis for graduate students, researchers and practitioners in the broad area of mathematical sciences and engineering, including operations research, economics, and finances. The focus is on problems with *constraints*, the analysis of which involves set-valued mappings and functions that are not differentiable, at least in the usual sense. A typical reader of the book should be familiar with multivariable calculus and linear algebra. Some basic knowledge in optimization, control, and elementary functional analysis is desirable but not necessary—all needed background material is included in the book.

The book is structured in 20 lectures, each one of which is devoted to a particular topic. The choice of topics reflects the author's views and interests and by no means covers all aspects of the field. The material is presented in a theorem-proof format, but there are also examples, exercises, and discussions. In each lecture, the proofs that are essential for understanding the concepts and techniques involved are given in the text, and reading these proofs is highly recommended. More technical or lengthy

e.g., as in the book by M. Morse, *Variational analysis: critical extremals and Sturmian extensions,* Interscience Publishers, 1973, which is mainly about calculus of variations at large.

proofs are given at the end of the lecture and could be skipped at a first reading. Most of the exercises are supplied with hints or detailed guides. Some classical results in analysis such as, e.g., (a version of) the Baire category theorem, are given without proofs in the lecture where they are first used. Each lecture has its own numbering of formulas; citing formulas from other lectures is avoided.

The book starts with a preparatory part presenting in a condensed form notations and terminology, along with some basic concepts and facts from functional analysis. Lecture 1 introduces standard optimization problems and in particular shows how to derive necessary optimality conditions in the format of generalized equations. Lecture 2 is about continuity of set-valued mappings. In addition, conditions are presented for continuity of the optimal value and optimal solution mappings of a general optimization problem depending on a parameter. Lipschitz continuity of set-valued mappings is also introduced in that lecture but explored more broadly in the following Lecture 3, which deals in particular with polyhedral mappings with application to linear programming, as well as outer Lipschitz continuity of piecewise polyhedral mappings.

The next several lectures are devoted to regularity properties of set-valued mappings with applications to nonlinear optimization. *Metric regularity* of mapping acting in metric spaces, together with the companion properties of Aubin continuity and linear openness, are introduced in Lecture 4. Lecture 5 is devoted to stability of metric regularity under perturbations as defined by the Lyusternik-Graves theorem. Lecture 6 presents the Robinson-Ursescu theorem, which gives a characterization of metric regularity of mappings with closed and convex graphs. Lecture 7 deals with derivative criteria for metric regularity that utilize generalized differentiation of set-valued mappings. *Strong regularity* is introduced in Lecture 8 together with the Robinson theorem, a far reaching extension of the classical implicit function theorem. The next one, Lecture 9, is devoted to two basic implicit function theorems for nondifferentiable functions, one due to F. Clarke and the other to B. Kummer. Lecture 10 characterizes strong regularity of mappings appearing in quadratic optimization problems in a Hilbert space setting, and in a basic nonlinear programming problem. *Strong subregularity* is introduced and discussed in Lecture 11. Lecture 12 presents a version of the Bartle-Graves theorem for set-valued mappings, which is concerned with the existence of *continuous selections* of the inverses of metrically regular mappings. Lecture 13 deals with so-called radius theorems associated with regularity properties. Lectures 7, 12, and 13 are technically involved and can be skipped at a first reading of the book; they are not used in the further lectures. Lecture 14 broadens the spectrum of applications of the regularity theory; it shows convergence of the Newton method applied to generalized equations involving mappings having the regularity properties given in the preceding lectures. A version of the Newton method for a class of nondifferentiable functions, the so-called semismooth functions, is presented in Lecture 15.

The final five lectures are devoted to applications of variational analysis to optimal control problems. The main focus is on the linear-quadratic optimal control problem with control constraints, which is relatively simple but still rich enough to allow demonstrating applications of basic ideas and techniques of variational analysis, without going deep into differential equations and functional analysis. Lecture 16 derives optimality conditions for the linear-quadratic problem. Lecture 17 considers a more general nonlinear optimal control problem and shows how to apply the theorems of Lyusternik-Graves, Robinson-Ursescu, and Robinson in order to obtain conditions for metric regularity and strong regularity of feasibility and optimality mappings for that problem. Lecture 18 is devoted to a discrete approximation of the linear-quadratic and a nonlinear optimal control problem, for which error estimates are obtained by using regularity properties of the mappings involved. In Lecture 19, optimal feedback control is introduced and discussed, and the existence of such a feedback is proven for the control-constrained linear-quadratic problem. The final Lecture 20 is devoted to model predictive control, which is presented as an approach to approximate optimal feedback; the main result presented is about the accuracy of such an approximation.

Parts of this book can be used in various graduate courses in the general area of optimization and control, depending on the audience and the specific goals of the lecturer. For example, combining the material of lectures 1–3, 8–11, and 15 with some adjustments and additions may become a one-semester course on modern optimization theory. Lectures 1–8, and 13–17 provide a basis for a course on variational analysis for mathematically mature students. Lectures 4–6, 12, and 15–20 may be used in a course on control, addressing in particular optimality, regularity, as well as more specific topics such as discrete approximations and model predictive control.

The idea to write this book was born in the spring of 2020, when I gave a one-semester graduate course on variational analysis in the Faculty of Mathematics and Informatics, Sofia University, Bulgaria. I was not able to find a suitable book that could be used as a textbook for the course. Then I prepared lecture notes which evolved into this book. While working on the book, new topics have been added to obtain a broader coverage of nonsmooth analysis and regularity of mappings, as well as topics in optimal control. The choice of the material in the book was also inspired by the decades long research collaboration of the author with students, postdocs, and faculty from the University of Michigan in Ann Arbor.

I am indebted to a number of people who have helped me with this book. During the preparation of the manuscript, I benefited from extensive discussions with V. Veliov, in particular concerning the lectures on control. Special thanks to R. Goebel, C. Josz, M. Krastanov, D. Liao-McPherson, J. Leung, Y. Stoev, and N. Zlateva, who had read preliminary drafts and made valuable suggestions, and especially to R. Rozenov, who also helped with the figures. I am also thankful to R. Bot, D. Drusvyatskiy, D. Klatte, I. Kolmanovsky, S. Robinson, and T. Zolezzi for their advice and encouragement. Many thanks to the anonymous referees for their valuable remarks. I obtained additional feedback from students who attended my course at Sofia University, and in particular from G. Angelov, S. Apostolov, M. Konstantinov, M. Nikolova, B. Stefanov, M. Tasheva, and I. Vasilev. The support from AFOSR under the grant FA9550-20-1-0385 is greatly appreciated.

This book would have never been written without the moral support of my family. In particular, I am thankful to my wife Dora for her patience and understanding, to my daughter Mira for her help with typesetting, and to my son Kiko for the many questions he asked me about this project.

Ann Arbor, MI, USA
June 2021

Asen L. Dontchev

In Memoriam. In finishing the Preface above, Asen Dontchev knew he had a terminal illness. However, in putting September 2021 beneath his name he had no inkling that, already on the 16th of that month, his life would come to its end.

Through his original research and many publications, he contributed hugely to the mathematics of optimization and control. This volume of lectures, brought forth in troublesome final circumstances, will help to spread the understanding of that subject and its evolving applications. It shows his remarkable qualities as a thinker and writer, being able to combine theoretical with practical while revealing the vital heart of every topic. He always stressed the importance of explaining things to newcomers to variational analysis, building on their diverse motivations from other areas and trying to ease their way around technicalities. For all this, he will be sadly missed and lastingly remembered.

Terry Rockafellar

Contents

Lecture 0
Notation, Terminology, and Some Functional Analysis

We start with a brief introduction to metric spaces. Let X be a set, $I\!R$ be the set of reals and $I\!R_+$ be the set of nonnegative reals. A *metric* ρ in X is a mapping $\rho : X \times X \to I\!R_+$ which has the following properties:

– $\rho(x, y) = 0 \quad \Longrightarrow \quad x = y$;

– $\rho(x, y) = \rho(y, x)$;

– $\rho(x, z) \le \rho(x, y) + \rho(y, z)$ (the triangle inequality).

The set X equipped with a metric ρ is called a *metric space* and denoted (X, ρ). In such a space the closed ball with center $x \in X$ and radius $r \in I\!R_+$ is the set

$$I\!B_r(x) = \left\{ y \in X \,\middle|\, \rho(y, x) \le r \right\}.$$

The open ball with center x and radius r is defined in the same way with $\le$ replaced by $<$ and denoted int $I\!B_r(x)$.

A metric space X is said to be *complete* when every Cauchy sequence $\{x_n\}$ in X converges. That is to say: if $\rho(x_n, x_m) \to 0$ as both n and m independently go to infinity, then there is $y \in X$ such that $\rho(x_n, y) \to 0$.

A *linear* or *vector* space is a set supplied with two operations: addition and multiplication by a scalar, which obey standard rules such as commutativity, associativity, and distributivity. All linear spaces considered are over the reals. A metric ρ acting in a linear space X is said to be *shift-invariant* when

$$\rho(y + z, y' + z) = \rho(y, y') \text{ for all } y, y', z \in X.$$

A linear space X is said to be *normed* when there is a mapping $\|\cdot\| : X \to I\!R$ defined on X and called *norm* that has the following properties:

– $\|x\| \ge 0$ for all $x \in X$ and $\|x\| = 0 \iff x = 0$;

– $\|\alpha x\| = |\alpha|\|x\|$ for all $x \in X$ and $\alpha \in I\!R$, where $|\alpha|$ denotes the absolute value of a real α: $|\alpha| = \max\{0, \alpha\}$;

– $\|x + y\| \le \|x\| + \|y\|$ for all $x, y \in X$.

© The Author(s), under exclusive license to Springer Nature Switzerland AG 2021

A. L. Dontchev, *Lectures on Variational Analysis*, Applied Mathematical Sciences 205, https://doi.org/10.1007/978-3-030-79911-3_0

A normed space is a metric space with a metric $\rho(x, y) = \|x - y\|$. In such a space the closed unit ball $I\!B_1(0)$ is denoted by $I\!B$ and the open unit ball is then int $I\!B$. A space that is linear, normed and complete is called a *Banach space* .

Given a linear space X, an *inner* or *scalar* product on X is a function $\langle \cdot, \cdot \rangle : X \times X \to I\!R$ with the following properties:

– $\langle x, y \rangle = \langle y, x \rangle$ for all $x, y \in X$;
– $\langle x_1 + x_2, y \rangle = \langle x_1, y \rangle + \langle x_2, y \rangle$ for all $x_1, x_2, y \in X$;
– $\langle \lambda x, y \rangle = \lambda \langle y, x \rangle$ for all $x, y \in X$ and $\lambda \in I\!R$;
– $\langle x, x \rangle \geq 0$ and $\langle x, x \rangle = 0 \iff x = 0$.

Then $\sqrt{\langle x, x \rangle}$ is a norm of x. A linear space X equipped with a scalar product and an associated norm which is complete with respect to that norm is said to be a *Hilbert space*. A standard fact for such a space is the Cauchy-Schwarz inequality

$$\langle u, v \rangle \leq \|u\|\|v\|.$$

The space consisting of all linear and continuous real-valued functions on a Banach space X is another Banach space which is *dual* or *adjoint* to X, denoted by X^*. The value that an $x^* \in X^*$ assigns to an $x \in X$ is called the duality mapping. The dual of the Banach space X^* is the bidual X^{**} of X. When every function $x^{**} \in X^{**}$ on X^* can be represented as $x^* \mapsto \langle x^*, x \rangle$ for some $x \in X$, the space X is called *reflexive*. In particular, when X is a Hilbert space with inner product $\langle x, y \rangle$, each $x^* \in X^*$ corresponds to a function $x \mapsto \langle x, y \rangle$ for some $y \in X$, so that X^* can be identified with X itself.

The n-dimensional Euclidean space, denoted $I\!R^n$, is a linear space of vectors

$$x = \begin{pmatrix} x_1 \\ \vdots \\ x_n \end{pmatrix}$$

equipped with the Euclidean norm

$$\|x\| = \sqrt{\langle x, x \rangle} = \left[\sum_{j=1}^{n} x_j^2 \right]^{1/2}$$

which is associated with the canonical inner product

$$\langle x, y \rangle = \sum_{j=1}^{n} x_j y_j.$$

The space $I\!R$ is the set of reals, the real line, whose norm is the absolute value $| \cdot |$.

Given a metric space (X, ρ), a point $x \in X$ and a set $C \subset X$, the quantity

$$d(x, C) = \inf_{y \in C} \rho(x, y)$$

is called the *distance* from x to C. We adopt the convention that the distance from any point to the empty set is $+\infty$. Any point y of C which is closest to x in the sense of achieving this distance is called a *projection* of x on C. The set of projections is denoted by $P_C(x)$.

In any metric space, a *neighborhood* of x is any set U for which there exists a positive number r such that $\mathbb{B}_r(x) \subset U$. We recall that the interior of a set $C \subset \mathbb{R}^n$ consists of all points x such that C is a neighborhood of x, whereas the closure of C consists of all points x such that the *complement* of C is *not* a neighborhood of x; C is *open* if it coincides with its interior and *closed* if it coincides with its closure. A nonempty set $C \subset X$ is closed if and only if every $x \in X$ with $d_C(x) = 0$ belongs to C. The interior is denoted by $\operatorname{int} C$. The union of *any* number of open sets is open, while the intersection of a *finite* number of open sets is open.

A set C is said to be *locally closed* at a point $x \in C$ when there exists a neighborhood U of x such that the intersection $C \cap U$ is a closed set. A set C is said to be *dense* in a closed set D when the closure $\operatorname{cl} C$ of C coincides with D, or equivalently, when for any $x \in D$ any neighborhood U of x contains elements of C. A set C is said to be *compact* when every open cover of it has a finite subcover. That is, C is compact if for every collection $\mathcal{T}$ of open sets $U \subset C$ such that $C \subset \cup_{\mathcal{T}} U$, there is a finite subset $\mathcal{F}$ of $\mathcal{T}$ such that $C \subset \cup_{\mathcal{F}} U$. In metric spaces this is the same as saying that every sequence $\{x_n\}$ with $x_n \in C$ has a subsequence which is convergent to an element of C. If $C \subset \mathbb{R}^n$ then compactness of C is equivalent to C being both bounded and closed.

A set C in a linear space X is said to be *convex* if for every $x, x' \in C$ and every $\lambda \in [0, 1]$, the point $\lambda x + (1 - \lambda)x'$ is in C. A function $f : X \to \mathbb{R}$ with domain containing a convex set C is said to be *convex* on C, or over C, or relative to C, if for every $x, x' \in C$ and every $\lambda \in (0, 1)$,

$$f(\lambda x + (1 - \lambda)x') \le \lambda f(x) + (1 - \lambda) f(x').$$

If $\le$ is replaced by $<$ for $x \ne x'$, the function f is said to be *strictly convex* on C. A function $f : X \to \mathbb{R}$ defined in a Banach space X is said to be *strongly convex* on C if there exists a constant $\alpha > 0$ such that the function $f(x) - \alpha\|x\|^2$ is convex. A function $f : X \to \mathbb{R}$ is *concave* when $-f$ is convex.

The *convex hull* of a set $C \subset \mathbb{R}^n$, which will be denoted by $\operatorname{co} C$, is the smallest convex set that includes C. It can be identified as the intersection of all convex sets that include C, but also can be described as consisting of all linear combinations $\lambda_0 x_0 + \lambda_1 x_1 + \cdots + \lambda_n x_n$ with $x_i \in C$, $\lambda_i \ge 0$, and $\lambda_0 + \lambda_1 + \cdots + \lambda_n = 1$; this is the Carathéodory theorem. The *closed convex hull* of C is the closure of the convex hull of C and denoted $\operatorname{cl} \operatorname{co} C$; it is the smallest closed convex set that contains C.

A sequence x_k with elements in a metric space (X, ρ) is said to strongly converge to x, or simply to converge to x, when for every $\varepsilon > 0$ there exists a natural number K such that for all $k \ge K$ we have $\rho(x_k, x) \le \varepsilon$. A sequence x_k in a Banach space X with a dual X^* is weakly convergent to x when for every $x^* \in X^*$ the sequence $\langle x^*, x_k \rangle$ converges to $\langle x^*, x \rangle$. For a Hilbert space X weak convergence means that for every $y \in X$ the sequence of reals $\langle y, x_k \rangle$ is convergent to $\langle y, x \rangle$. Weak convergence determines weak compactness, i.e., a set C in a Hilbert space X is weakly compact when every sequence has a weakly convergent subsequence. A basic result in that context says that any bounded, closed and convex set in a Hilbert space is weakly compact.

In further lines we introduce notations and terminology for mappings. A mapping F acting from a set X to a set Y is generally denoted as

$$X \ni x \mapsto F(x) \subset Y,$$

where $x \in X$ and $F(x)$ is the image of x which in general is a subset of Y. The *domain* of a mapping F acting between X and Y is

$$\text{dom } F = \{x \in X \mid F(x) \neq \emptyset\},$$

the *graph* of F is

$$\text{gph } F = \{(x, y) \in X \times Y \mid y \in F(x)\},$$

while the *range* of F is

$$\text{rge } F = \{y \in Y \mid \text{ there exists } x \in \text{dom } F \text{ with } y \in F(x)\}.$$

A mapping $F : X \to Y$ is said to be a *function* when for every $x \in X$ the image $F(x) \subset Y$ is just one point or the empty set. We will also consider mappings for which the image $F(x)$ of a point x may consist of more than one point and call such mappings *set-valued mappings*. In the notation for set-valued mappings we use capital letters and double arrows, e.g., $F : X \rightrightarrows Y$, versus small letters and single arrows for functions, e.g., $f : X \to Y$. Every function may be viewed as a set-valued mapping. A set-valued mapping which is not a function, that is, having multiple values at certain points in its domain, is said to be a *multivalued mapping*, in contrast to a function, which is a *single-valued mapping* in its domain.

We define the *inverse* of a set-valued mapping $F : X \rightrightarrows Y$ as

$$Y \ni y \mapsto F^{-1}(y) = \{x \in X \mid y \in F(x)\}.$$

According to this definition, every mapping has an inverse. In particular, the inverse of a function always exists, but it may be multivalued, that is, not a function. A simple example is given in Fig. 0.1.

We introduce next a concept which identifies the case when a set-valued mapping is locally a function . Let X and Y be metric spaces.

Single-Valued Graphical Localization of a Set-Valued Mapping. *For $F : X \rightrightarrows Y$ and a pair $(\bar{x}, \bar{y}) \in \text{gph } F$, a function s is said to be a* single-valued graphical localization *of F around $\bar{x}$ for $\bar{y}$ if there exist neighborhoods U of $\bar{x}$ and V of $\bar{y}$ such that $U \subset \text{dom } s$ and*

$$\text{gph } s = (U \times V) \cap \text{gph } F,$$

so that

$$s : x \mapsto \begin{cases} F(x) \cap V & \text{when } x \in U, \\ \emptyset & \text{otherwise.} \end{cases}$$

In the example displayed in Fig. 0.1, the inverse of the function $x \mapsto x^2$ has a single-valued localization around any $x > 0$ for $y = \sqrt{x}$ and another one around any $x > 0$ for $y = -\sqrt{x}$; moreover, it has no single-valued localization around 0 for 0 and is empty valued for any $x < 0$.

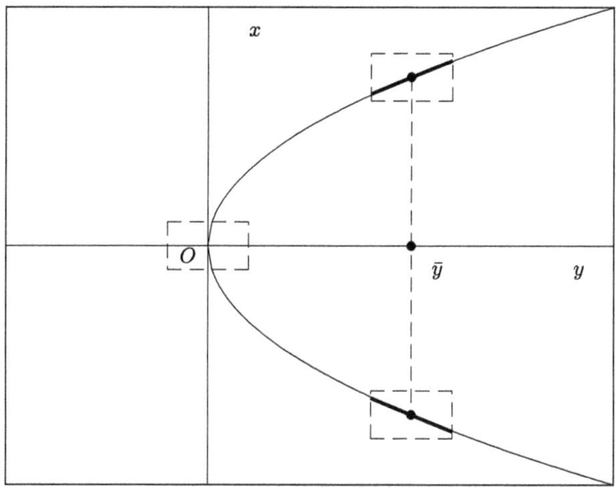

Fig. 0.1: The inverse of the function $x \mapsto x^2$

Let X and Y be Banach spaces. A *linear* mapping A acting from X to Y is a function with dom $A = X$ which obeys the rule for linearity:

$$A(\alpha x + \beta y) = \alpha Ax + \beta Ay \ \text{ for all } \ x, \ y \in X \ \text{ and all scalars } \ \alpha, \beta \in I\!R.$$

A linear mapping $A : X \to Y$ acting between Banach spaces X and Y is *bounded* when there exists a constant $\alpha \geq 0$ such that

$$||Ax|| \leq \alpha ||x|| \quad \text{for all } \ x \in X.$$

The space of linear bounded mappings acting from a Banach space X to a Banach space Y is denoted by $\mathcal{L}(X, Y)$. It is a Banach space when equipped with the operator norm $\sup_{x \in I\!B} ||Ax||$. A mapping $A \in \mathcal{L}(X, Y)$ is said to be *surjective* or *onto* when for every $y \in Y$ there is a $x \in X$ with $Ax = y$; this is denoted as $AX = Y$. We also use the notation ker $A = \{x \in X \mid Ax = 0\}$.

Let A be a linear and bounded mapping A acting from Hilbert space X into itself. Then the mapping A^* defined as

$$\langle A^* x, y \rangle = \langle Ay, x \rangle \quad \text{for all } \ x, y \in X$$

is linear and bounded and is called the *adjoint mapping* associated with A. If $A = A^*$ the mapping A is said to be *selfadjoint*.

In $I\!R^n$ we make a distinction between a linear mapping and its matrix. Specifically, a linear mapping $A : I\!R^n \to I\!R^m$ is represented by a matrix, which, with some abuse of notation, we denote again by A; here the matrix A is with m rows, n columns, and components $a_{i,j}$. The transpose of A, denoted by A^T, represents the adjoint of the mapping A. If the matrix A is nonsingular, which requires $m = n$, then the inverse A^{-1} of A is also a linear mapping represented by the inverse of the matrix A. More generally, if $m \leq n$ and the rows of the matrix A are linearly independent, then the rank of the matrix A is m and the mapping A is surjective. In this case the matrix AA^T is nonsingular. On the other hand, if $m \geq n$ and the columns of A are linearly independent then $A^\mathsf{T}A$ is nonsingular. Both the identity mapping and its matrix will be denoted by I, regardless of dimensionality.

Let (X, ρ) and (Y, ρ) be metric spaces. A function $f : X \to Y$ is said to be *continuous* at $\bar{x} \in \operatorname{dom} f$ if for every $\varepsilon > 0$ there exists $\delta > 0$ such that $\rho(f(x), f(\bar{x})) \leq \varepsilon$ whenever $\rho(x, \bar{x}) \leq \delta$. A function f is said to be continuous on, or over, or relative to, a set $D \subset \operatorname{dom} f$ if it is continuous at every $x \in D$. A function $f : X \to Y$ is said to be *open* at $\bar{x} \in \operatorname{dom} f$ when for any neighborhood U of $\bar{x}$ the set $f(U)$ is a neighborhood of $f(\bar{x})$. A function $f : X \to I\!R$ is said to be *lower semicontinuous* on a set C if for every point $x \in C$ and every sequence $x_k \in C$ convergent to x one has $\liminf_{k\to\infty} f(x_k) \geq f(x)$. Symmetrically, a function $f : X \to I\!R$ is *upper semicontinuous* on a set C if for every point $x \in C$ and every sequence $x_k \in C$ convergent to x one has $\limsup_{k\to\infty} f(x_k) \leq f(x)$. Recall the classical Weierstrass theorem:

Weierstrass Theorem. *Let C be a nonempty compact set in a metric space X. Then every lower semicontinuous function on C attains its minimum on C and every upper semicontinuous function on C attains its maximum on C. Hence, any continuous on C function attains both its minimum and maximum on C.*

A function $f : X \to I\!R$ acting on a Hilbert space X is said to be weakly lower semicontinuous at a point x if it is lower semicontinuous with respect to the weak convergence in X. The Weierstrass theorem extends to the weak versions of compactness and lower semicontinuity: any weakly lower semicontinuous function f on Hilbert space X attains its minimum on a weakly compact set and, in particular, on a closed and convex set. A continuous function which is convex, is weakly lower semicontinuous.

Let X and Y be Banach spaces. A function $f : X \to Y$ is said to be *Fréchet differentiable* at a point x when $x \in \operatorname{int} \operatorname{dom} f$ and there is a linear and bounded mapping $A : X \to Y$ with the property that for every $\varepsilon > 0$ there exists $\delta > 0$ such that

$$\|f(x + h) - f(x) - Ah\| \leq \varepsilon\|h\| \quad \text{for every } h \in X \text{ with } \|h\| < \delta.$$

If such a mapping A exists, it is unique; it is denoted by $Df(x)$ and called the *Fréchet derivative* of f at x. A function $f : X \to Y$ is said to be *twice Fréchet differentiable* at a point $x \in \operatorname{int} \operatorname{dom} f$ when it is Fréchet differentiable at x and there is a bilinear mapping $N : X \times X \to Y$ with the property that for every $\varepsilon > 0$ there exists $\delta > 0$

with

$$\|f(x+h) - f(x) - Df(x)h - N(h,h)\| \le \varepsilon\|h\|^2 \quad \text{for every } h \in X \text{ with } \|h\| < \delta.$$

If such a mapping N exists, it is unique and is called the *second Fréchet derivative* of f at x, denoted by $D^2 f(x)$. Higher-order derivatives can be defined accordingly. (For functions acting in finite dimensions these are the usual derivatives from calculus; then we omit Fréchet.) When the Fréchet derivative mapping $x \mapsto Df(x)$ exists and is continuous (with respect to the operator norm) on an open set $C \subset X$, then we say that the function f is *continuously differentiable* on C; we also call such a function *smooth* on C. Analogously, for an integer k we define k times continuously differentiable functions. The set of such functions is denoted by C^k.

For a function $f : P \times X \to Y$ and a pair $(p, x) \in \operatorname{int} \operatorname{dom} f$, the *partial Fréchet derivative* mapping $D_x f(p, x) : X \to Y$ of f with respect to x at (p, x) is the Fréchet derivative of the function $g(y) = f(p, y)$ at x. If the partial derivative mapping is a continuous function in a neighborhood of (p, x), then f is said to be continuously differentiable with respect to x around (p, x).

For a function $f : I\!R^n \to I\!R^m$ we distinguish between the derivative as a linear mapping and its matrix. The $m \times n$ matrix that represents the derivative $Df(x)$ at x is called the *Jacobian* of f at x and is denoted by $\nabla f(x)$. The second derivative is denoted by $\nabla^2 f(x)$ and so on. In the notation $x = (x_1, \ldots, x_n)$ and $f = (f_1, \ldots, f_m)$, the components of $\nabla f(x)$ are the partial derivatives of the component functions f_i:

$$\nabla f(x) = \left(\frac{\partial f_i}{\partial x_j}(x) \right)_{i,j=1}^{m,n}.$$

For $f : I\!R^d \times I\!R^n \to I\!R^m$ the partial derivative $\nabla_x f(p, x)$ is represented by an $m \times n$ matrix, denoted $\nabla_x f(p, x)$ and called the partial Jacobian. It's a standard fact from calculus that if the function $(p, x) \mapsto f(p, x)$ is differentiable with respect to both p and x around $(\bar{p}, \bar{x})$ and the partial Jacobian mappings $(p, x) \mapsto \nabla_x f(p, x)$ and $(p, x) \mapsto \nabla_p f(p, x)$ are continuous around $(\bar{p}, \bar{x})$, then f is continuously differentiable around $(\bar{p}, \bar{x})$.

In this book we will also employ the following weaker notion of derivative; for simplicity we stick with finite dimensions. For $f : I\!R^n \to I\!R^m$, a point $\bar{x} \in \operatorname{dom} f$ and a vector $d \in I\!R^n$, the limit

$$f'(\bar{x}; d) = \lim_{t \searrow 0} \frac{f(\bar{x} + td) - f(\bar{x})}{t},$$

when it exists, is called the *directional derivative* of f at $\bar{x}$ for d. If this directional derivative exists for every d, f is said to be *directionally differentiable* at $\bar{x}$.

Let X and Y be metric spaces where both metrics are denoted by ρ but may be different. A function $f : X \to Y$ is said to be *Lipschitz continuous* relative to a set C, or on a set C, if $C \subset \operatorname{dom} f$ and there is a constant $\kappa \ge 0$ such that

$$\rho(f(x'), f(x'')) \le \kappa\rho(x', x'') \quad \text{for all } x', x'' \in C. \tag{1}$$

If f is Lipschitz continuous relative to a neighborhood of a point $x \in \operatorname{int} \operatorname{dom} f$, then f is said to be Lipschitz continuous *around* x. We say further, in the case of an open set C, that f is locally Lipschitz continuous on C if it is a Lipschitz continuous function around every point x of C.

For a function $f : X \to Y$ and a point $x \in \operatorname{int} \operatorname{dom} f$, the *Lipschitz modulus* of f at x, denoted $\operatorname{lip}(f; x)$, is the infimum of the set of values of κ for which there exists a neighborhood C of x such that (1) holds. Equivalently,

$$\operatorname{lip}(f; x) = \limsup_{\substack{x',x''\to x,\\ x'\neq x''}} \frac{\rho(f(x'), f(x''))}{\rho(x', x'')}.$$

A function f is Lipschitz continuous around x if and only if $\operatorname{lip}(f; x) < \infty$. For an open set C, a function f is *locally Lipschitz continuous* on C exactly when $\operatorname{lip}(f; x) < \infty$ for every $x \in C$.

Let P, X and Y be metric spaces. A function $f : P \times X \to Y$ is said to be Lipschitz continuous with respect to x around $(p, x) \in \operatorname{int} \operatorname{dom} f$ when the function $y \mapsto f(p, y)$ is Lipschitz continuous around x; the associated Lipschitz modulus of f with respect to x is denoted by $\operatorname{lip}_x(f; (p, x))$. We say f is Lipschitz continuous with respect to x *uniformly in* p around $(p, x) \in \operatorname{int} \operatorname{dom} f$ when there are neighborhoods Q of p and U of x along with a constant κ and such that

$$\rho(f(p', x''), f(p', x')) \leq \kappa\rho(x'', x') \quad \text{for all } x'', x' \in U \text{ and } p' \in Q.$$

Accordingly, the partial uniform Lipschitz modulus with respect to x has the form

$$\widehat{\operatorname{lip}}_x(f; (p, x)) := \limsup_{\substack{x'',x'\to x,p'\to p,\\ x''\neq x'}} \frac{\rho(f(p', x''), f(p', x'))}{\rho(x'', x')}.$$

A one-point version of the Lipschitz continuity is a property called *calmness*. A function $f : X \to Y$ is said to be *calm* at x relative to a set D in X if $x \in D \cap \operatorname{dom} f$ and there exists a constant $\kappa \geq 0$ such that

$$\rho(f(x'), f(x)) \leq \kappa\rho(x', x) \quad \text{for all } x' \in D \cap \operatorname{dom} f. \tag{2}$$

For a function $f : X \to Y$ and a point $x \in \operatorname{dom} f$, the calmness modulus of f at x, denoted $\operatorname{clm}(f; x)$, is the infimum of the set of values $\kappa \geq 0$ for which there exists a neighborhood D of x such that (2) holds. The definition of the partial uniform calmness modulus is completely analogous to that of the partial uniform Lipschitz modulus.

Let X and Y be Banach spaces. Having the concept of calmness, we can interpret the Fréchet differentiability of a function $f : X \to Y$ at a point $x \in \operatorname{int} \operatorname{dom} f$ as the existence of a linear mapping $Df(x) : X \to Y$ such that

$$\operatorname{clm}(e; x) = 0 \quad \text{for } e(x') = f(x') - [f(x) + Df(x)(x' - x)].$$

Furthermore, we have that $\operatorname{clm}(f;x) = \|Df(x)\|$. A sharper concept of derivative is tied up with the Lipschitz modulus. A function $f : X \to Y$ is said to be *strictly* Fréchet differentiable at a point x if there is a linear and bounded mapping $A : X \to Y$ such that

$$\operatorname{lip}(e;x) = 0 \quad \text{for } e(x') = f(x') - [f(x) + A(x' - x)].$$

Specifically, in this case we have that for each $\varepsilon > 0$ there exists a neighborhood U of x such that

$$\|f(x'') - [f(x') + Df(x)(x'' - x')]\| \le \varepsilon\|x'' - x'\| \quad \text{for every } x'', x' \in U.$$

Clearly, the strictly Fréchet differentiable functions are Fréchet differentiable and the continuously Fréchet differentiable functions are strictly Fréchet differentiable functions, with $\operatorname{lip}(f;x) = \|Df(x)\|$. Sometimes, when clear from the context, we omit "Fréchet."

We introduce next the notion of semidifferentiability. First, we need the following definition: A function $\varphi : X \to Y$ is said to be *positively homogeneous* if $0 = \varphi(0)$ and $\varphi(\lambda w) = \lambda\varphi(w)$ for all $w \in \operatorname{dom}\varphi$ and $\lambda > 0$. This mean geometrically that the graph of φ is a cone in $X \times Y$. Any linear function is positively homogeneous in particular.

Semidifferentiability. *A function $f : X \to Y$ is said to be semidifferentiable at x if there exists a continuous and positively homogeneous function $\varphi : X \to Y$ such that*

$$\operatorname{clm}(e;x) = 0 \quad \textit{for } e(x') = f(x') + \varphi(x' - x).$$

If the stronger condition holds that

$$\operatorname{lip}(e;x) = 0 \quad \textit{for } e(x') = f(x') + \varphi(x' - x),$$

then f is said to be strictly semidifferentiable at x. Either way, the function φ, necessarily unique, is called the semiderivative of f at x and denoted by $Df(x)$. In the literature, this kind of derivative is also called Bouligand or B-derivative.

For a function f and a point $x \in \operatorname{dom} f$, semidifferentiability of f at x is equivalent to the existence of the limit

$$\lim_{t \searrow 0, w' \to w} \frac{f(x + tw') - f(x)}{t} \quad \text{for every } w \in X\ .$$

If a function f is semidifferentiable at x, then f is in particular directionally differentiable at x and has

$$f'(x;w) = Df(x)(w) \quad \text{for all } w.$$

When $\operatorname{lip}(f;x) < \infty$, directional differentiability at x in turn implies semidifferentiability at x.

When the semiderivative $Df(x)$ is linear, semidifferentiability turns into differentiability, and strict semidifferentiability turns into strict differentiability. The connections known between $Df(x)$ and the calmness modulus and Lipschitz modulus of f at x under differentiability can be extended to semidifferentiability by adopting the definition that

$$\|\varphi\| = \sup_{\|x\|\leq 1} \|\varphi(x)\| \quad \text{for a positively homogeneous function } \varphi.$$

We then have $\operatorname{clm}(Df(x); 0) = \|Df(x)\|$ and consequently $\operatorname{clm}(f; x) = \|Df(x)\|$, which in the case of strict semidifferentiability becomes $\operatorname{lip}(f; x) = \|Df(x)\|$.

Examples.

(1) The function $f(x) = e^{|x|}$ for $x \in I\!R$ is not differentiable at 0, but it is semidifferentiable there and its semiderivative is given by $Df(0) : w \mapsto |w|$. This is actually a case of strict semidifferentiability. Away from 0, f is of course continuously differentiable (hence strictly differentiable).

(2) The function $f(x_1, x_2) = \min\{x_1, x_2\}$ on $I\!R^2$ is continuously differentiable at every point away from the line where $x_1 = x_2$. On that line, f is strictly semidifferentiable with

$$Df(x_1, x_2)(w_1, w_2) = \min\{w_1, w_2\}.$$

(3) A function of the form $f(x) = \max\{f_1(x), f_2(x)\}$, with f_1 and f_2 continuously differentiable from $I\!R^n$ to $I\!R$, is strictly differentiable at all points x where $f_1(x) \neq f_2(x)$ and semidifferentiable where $f_1(x) = f_2(x)$, the semiderivative being given there by

$$Df(x)(w) = \max\{Df_1(x)(w), Df_2(x)(w)\}.$$

However, f might not be strictly semidifferentiable at such points.

The semiderivative obeys standard calculus rules, such as semidifferentiation of a sum, product and ratio, and, most importantly, the chain rule. Here are some further properties of the semiderivative: Let f be semidifferentiable at x and let g be Lipschitz continuous and semidifferentiable at $y := f(x)$. Then the composition $g \circ f$ is semidifferentiable at x and

$$D(g \circ f)(x) = Dg(y) \circ Df(x).$$

Let f be strictly semidifferentiable at x and g be strictly differentiable at $f(x)$. Then $g \circ f$ is strictly semidifferentiable at x.

Lecture 1
Basics in Optimization

An *optimization problem* is typically a problem of finding minimum or maximum of a real-valued function f relative to a set C. The function f is called the *objective* or *cost* function, while the set C over which the minimization or maximization takes place is called the *feasible set* usually given by *constraints*. The problem of minimizing f over C consists of finding an element $\bar{x}$ in C such that

$$f(\bar{x}) \leq f(x) \quad \text{for all} \quad x \in C.$$

This problem is written as

$$\min f(x) \;\; \text{subject to} \;\; x \in C \qquad \text{or} \qquad \min_{x \in C} f(x).$$

Stated in that way, this is a problem of finding a *global* minimum of f over C. A point $\bar{x}$ is a *local* minimum of f over C if there exists a neighborhood U of $\bar{x}$ such that $f(\bar{x}) \leq f(x)$ for all $x \in C \cap U$. Clearly, every global minimum is also a local minimum but the converse is not true. A maximization problem consists of finding $\bar{x} \in C$ such that $f(\bar{x}) \geq f(x)$ for all $x \in C$ and can be stated equivalently as a minimization problem, inasmuch as

$$\max_{x \in C} f(x) = -\min_{x \in C}(-f(x)).$$

In this book we will consider mainly two kinds of optimization problems. First come *mathematical programming* problems, where the feasible set is a subset of an Euclidean space usually given by equalities and inequalities. The name "programming" most likely stems from the time when optimization problems were solved on early computers; it propagated to problem classes such as linear programming, quadratic programming, convex programming, nonlinear programming, etc., where linear, quadratic, convex, and nonlinear correspond to the type of functions involved in the objective function and the constraints. Then we will focus on *optimal control* problems, where the feasible set is a set of functions in an infinite-dimensional

© The Author(s), under exclusive license to Springer Nature Switzerland AG 2021
A. L. Dontchev, *Lectures on Variational Analysis*, Applied Mathematical Sciences 205, https://doi.org/10.1007/978-3-030-79911-3_1

space and the constraints involve differential equations describing the evolution of a control system.

A minimum of a function f over a set C may not exist, in which case one may ask what is the infimum (the exact lower bound) of f over C denoted as $\inf_{x\in C} f(x)$. That infimum, which in particular is a minimum if attained, is said to be the *optimal value*. The usual convention is that infimum over an empty set is $+\infty$, in which case it makes sense to assume that the feasible set C is a subset of dom f or at least has a nonempty intersection with it. A sequence $x^k \in C$ is said to be *minimizing* when $f(x^k)$ converges to the optimal value as $k \to \infty$. According to a classical theorem by Weierstrass given in the preparatory section, a problem of (global) minimum always has a solution provided that the feasible set C is nonempty and compact, and the objective function f is lower semicontinuous over C.

In the preparatory section we defined convex sets and functions. Suppose that X is a Banach space and an objective function $f : X \to I\!R$ has a local minimum $\bar{x}$ over a convex set C in X. In addition, let f be Fréchet differentiable at $\bar{x}$ with derivative $Df(\bar{x})$. Denote by $\langle\cdot,\cdot\rangle$ the duality mapping between X and its dual X^*. Of course, for $X = I\!R^n$ this is the usual scalar product. Let x be any point in C. Then, from the convexity of C it follows that $\bar{x} + t(x - \bar{x}) \in C$ for all $t \in [0, 1]$. Hence, from the assumption for local optimality, we obtain $f(\bar{x} + t(x - \bar{x})) \geq f(\bar{x})$ for all sufficiently small $t > 0$. Utilizing the differentiability of f at $\bar{x}$, we obtain

$$0 \leq \frac{f(\bar{x} + t(x - \bar{x})) - f(\bar{x})}{t} = \langle Df(\bar{x}), x - \bar{x}\rangle + \frac{o(t)}{t},$$

where $o(t)$ here and later denotes an expression such that $o(t)/t \to 0$ as $t \searrow 0$. Passing to the limit with $t \searrow 0$ and having in mind that x is an arbitrary point in C, we obtain

$$\langle Df(\bar{x}), x - \bar{x}\rangle \geq 0 \quad \text{for all} \quad x \in C. \tag{1}$$

Thus, inasmuch as the first derivative is involved, we obtain (1) as a *first-order necessary condition* for local optimality of $\bar{x}$. When $C = X$, condition (1) becomes

$$Df(\bar{x}) = 0,$$

which is the standard zero derivative test for a local minimum (and also maximum) known as the *Fermat rule*.

Every solution of the optimality condition (1) is said to be a *stationary point* for f on C. The set of local minima of f over C is a subset of the set of stationary points. For a convex and closed set C and a function $f : X \to I\!R$ that is convex relative to C, every local minimum of f over C is a global minimum as well. Moreover, the first-order necessary optimality condition (1) becomes also sufficient; that is, $\bar{x}$ is a global minimum of f over C if and only if condition (1) is satisfied. Equivalently, the set of stationary points of f on C coincides with the set of global minima of f over C. Both statements follow from the property of a convex function that

$$f(x) \geq f(\bar{x}) + \langle Df(\bar{x}), x - \bar{x}\rangle \quad \text{for every} \quad x \in C.$$

Indeed, using (1) in the last inequality, we get that $f(x) \geq f(\bar{x})$ for all $x \in C$.

Condition (1) is a special case of a variational inequality problem defined next:

Variational Inequality. *Let X be a Banach space and X^* be its dual, with a duality mapping $\langle \cdot, \cdot \rangle$. Given a function $g : X \to X^*$ and a set $C \subset X$, the problem*

$$\text{find } x \in C \text{ such that } \langle g(x), u - x \rangle \geq 0 \text{ for all } u \in C \tag{2}$$

is said to be a variational inequality problem.

In some cases it is convenient to cast variational inequalities in terms of *normal cones*. Recall that a set K is a *cone* if $v \in K$ implies $\lambda v \in K$ for every $\lambda \geq 0$. In this book we mainly use the following definition of a normal cone to a closed convex set:

Normal Cone to a Closed Convex Set. *Let C be a closed and convex set. For a point $x \in C$, the set*

$$N_C(x) = \{y \in X^* \mid \langle y, u - x \rangle \leq 0 \text{ for all } u \in C\}$$

is called the normal cone to C at the point x.

For any closed and convex set C and for any $x \in C$, the normal cone $N_C(x)$ is a closed and convex cone. Sometimes it is convenient to define the normal cone outside C as $N_C(x) = \emptyset$ for $x \notin C$. Then the domain of the *normal cone mapping* $x \mapsto N_C(x)$ is the set C.

Example. Let C be the positive orthant in $I\!R^2$; that is, $C = \{x = (x_1, x_2) \in I\!R^2 \mid x_1 \geq 0, x_2 \geq 0\}$. Then the normal cone $N_C(x)$ depends on the location of the point $x = (x_1, x_2)$. If x is in the interior of C, then the normal cone $N_C(x) = \{0\}$. If $x \in C$ is such that $x_1 > 0, x_2 = 0$, then $N_C(x) = \{x \in I\!R^2 \mid x_1 = 0, x_2 \leq 0\}$. Symmetrically, $x \in C$ satisfies $x_1 = 0, x_2 > 0$, then $N_C(x) = \{x \in I\!R^2 \mid x_1 \leq 0, x_2 = 0\}$. Finally, if $x \in C$ of the origin $(0, 0)$, then N_C is the negative orthant in $I\!R^2$, $N_C(x) = \{x \in I\!R^2 \mid x_1 \leq 0, x_2 \leq 0\}$.

In terms of the normal cone, the variational inequality (2) can be written as the inclusion

$$g(x) + N_C(x) \ni 0. \tag{3}$$

Noting that the optimality condition (1) is a variational inequality of a special form, the following statement is a necessary condition for optimality in terms of the normal cone mapping: if $\bar{x}$ is a local minimum of a function $f : X \to I\!R$ over a closed and convex set C in a Banach space X and f is Fréchet differentiable at $\bar{x}$, then the following condition holds:

$$Df(\bar{x}) + N_C(\bar{x}) \ni 0. \tag{4}$$

If in addition f is convex over C, then condition (4) becomes a necessary and sufficient condition for a global minimum. Furthermore, if f is strictly convex, then any (global) minimum is unique. Indeed, if we assume that f has two different minima x_1 and x_2 over C, then the point $x = (x_1 + x_2)/2$ must satisfy $f(x) < f(x_1)/2 + f(x_2)/2 = \min_{x \in C} f(x)$, which contradicts the definition of a minimum.

The variational inequalities are particular cases of *generalized equations*, a name used in this book for inclusions of the form

$$0 \in f(x) + F(x),$$

where f is a function and F is a set-valued mapping acting, say, between Banach spaces X and Y. The model of a generalized equation covers a large territory. The case of nonlinear equations corresponds to having $F = 0$, whereas by $Fv - K$ for a fixed set K one gets various constraint systems. For example, if $Y = I\!R^m$ and $K = I\!R^m_+$, we obtain the system of inequalities $f_i(x) \leq 0$ for $j = 1, \ldots, m$. When F is the normal cone mapping N_C associated with a closed, convex set C, we have a variational inequality. Necessary optimality conditions to optimization problems are represented by variational inequalities.

Let $f : X \to I\!R$ be a convex function on a Banach space X and let $\bar{x} \in \operatorname{dom} f$. An element x^* of the dual space X^* is said to be a *subgradient* of f at $\bar{x}$ when

$$f(x) - f(\bar{x}) \geq \langle x^*, x - \bar{x} \rangle \quad \text{for all } x \in X.$$

The set of all subgradients of f at $\bar{x}$ is the *subdifferential* of f at $\bar{x}$ denoted by $\partial f(\bar{x})$; that is,

$$\partial f(\bar{x}) = \{x^* \in X^* \mid f(\bar{x} + h) \geq f(x) + \langle x^*, h \rangle \text{ for all } h \in X\}.$$

If f is continuously differentiable around $\bar{x}$, then its subdifferential at $\bar{x}$ consists of one element, the derivative $Df(\bar{x})$. Directly from the definition of a global minimum we obtain the following specification of the Fermat rule: $\bar{x}$ is a global minimum of f if and only if $0 \in \partial f(\bar{x})$.

Of particular interest in finite-dimensional optimization are the polyhedral sets.

Polyhedral Set. *A set $C \subset I\!R^n$ is said to be polyhedral when there exist a natural number m, m vectors $b_i \in I\!R^n$, and m reals a_i such that*

$$C = \{x \in I\!R^n \mid \langle b_i, x \rangle \leq a_i,\ i = 1, \ldots, m\}. \tag{5}$$

If $a_i = 0$ for all i in (5)*, then C is a polyhedral cone.*

The polyhedral sets are closed and convex. Given a polyhedral set C described as in (5) and a point $x \in C$, the set of *active* constraints at x is defined as

$$I(x) = \{i \mid \langle b_i, x \rangle = a_i\}.$$

The normal cone to a polyhedral set of the form (5) has the following representation:

$$N_C(x) = \{v \in I\!R^n \mid v = \sum_{i=1}^{m} y_i b_i \text{ with } y_i \geq 0,\ i \in I(x),\ \ y_i = 0,\ i \notin I(x)\}.$$

Polar Cone. *For a closed, convex cone K in $I\!R^n$, its* polar K^* *is defined by*

$$K^* = \{ y \,|\, \langle x, y \rangle \le 0 \;\; \text{for all } x \in K \}.$$

The polar K^* to a closed, convex cone K is likewise a closed, convex cone, and its polar $(K^*)^*$ is in turn K. Furthermore, the normal vectors to K and K^* are related by

$$y \in N_K(x) \iff x \in N_{K^*}(y) \iff x \in K,\; y \in K^*,\; \langle x, y \rangle = 0.$$

Normal cones to convex and closed sets are related through polarity to tangent cones, which we introduce next (Fig. 1.1):

Tangent Cone. *For a closed set $C \subset X$ and a point $\bar{x} \in C$, a vector w is said to be* tangent *to C at $\bar{x}$ if there exists a sequence $x_k \in C$, $x_k \to \bar{x}$, along with a sequence of scalars $\tau_k \searrow 0$ such that $w_k = (x_k - \bar{x})/\tau_k \to w$. The set of all vectors w that are tangent to C at a point $\bar{x} \in C$ is called the tangent cone to C at $\bar{x}$ and denoted as $T_C(\bar{x})$.*

That the tangent cone is a closed cone that can be shown directly from the definition. If $\bar{x}$ is in the interior of C, then $T_C(\bar{x}) = X$. If C is a singleton, say $C = \{\bar{x}\}$, then $T_C(\bar{x}) = \{0\}$. For a polyhedral set C as in (5) and a point $\bar{x} \in C$ the tangent cone $T_C(\bar{x})$ is described in terms of the set of active constraints $I(x)$:

$$T_C(x) = \{w \in I\!R^n \mid \langle b_i, w \rangle \le 0 \;\; \text{for} \;\; i \in I(x)\}.$$

The tangent cone $T_C(x)$ to a closed, convex set C at a point $x \in C$ is polar to the normal cone $N_C(x)$: specifically, one has

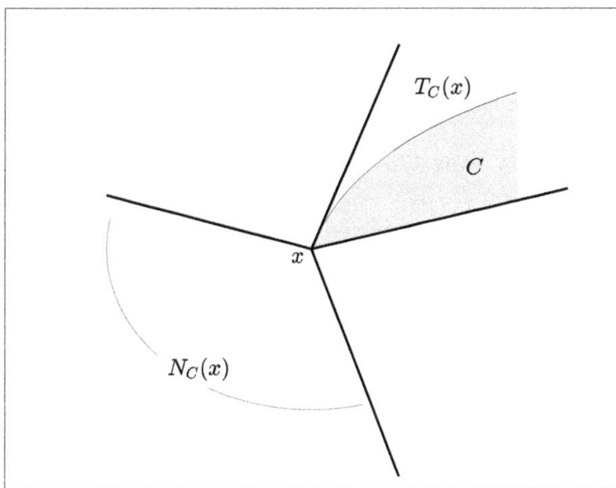

Fig. 1.1: Tangent and normal cones to a convex set

$$T_C(x) = N_C(x)^*, \qquad N_C(x) = T_C(x)^*.$$

We can state a necessary condition for a local minimum over a closed set C in terms of a tangent cone. Let $\bar{x}$ be a local minimum of a function f over a closed set C and let $w \in T_C(\bar{x})$. By definition, there exist sequences $x_k \to \bar{x}$ in C and $\tau_k \searrow 0$ such that $(x_k - \bar{x})/\tau_k \to w$. By repeating the argument before (1) we obtain that

$$Df(\bar{x})w \geq 0 \quad \text{for all} \;\; w \in T_C(\bar{x}).$$

A basic example of an optimization problem is the problem of projection. Let us briefly review the projection problem in $I\!R^n$. Recall that $\bar{x}$ is a projection of a point y over a set C if it is a solution of the minimization problem with the objective function $x \mapsto \|x - y\|$ and the feasible set C. The optimal value of this problem is the value $d(y, C)$ of the distance function $y \mapsto d(y, C)$, and the solutions set is the set of projections $P_C(y)$. For a nonempty convex set $C \subset I\!R^n$, the distance function is Lipschitz continuous on $I\!R^n$ with Lipschitz constant 1. Furthermore, the projection set $P_C(y)$ is nonempty, closed, and bounded for every $x \in I\!R^n$. If C is convex and closed, the projection mapping P_C is single-valued, a function, acting from $I\!R^n$ onto C, which moreover is Lipschitz continuous with Lipschitz constant 1 but may be not differentiable everywhere. For a closed and convex set C, there is a simple connection between the normal cone mapping N_C and the projection mapping P_C, as follows:

$$v \in N_C(x) \iff P_C(x + v) = x.$$

For the mappings N_C and P_C this implies that

$$N_C = P_C^{-1} - I, \qquad P_C = (I + N_C)^{-1}.$$

From the latter we obtain that the variational inequality (2) can be written as an equation, namely

$$g(x) + N_C(x) \ni 0 \iff P_C(x - g(x)) - x = 0.$$

First-Order Optimality over Polyhedral Sets. Let C be a polyhedral set of the form (5). Given a function $f : I\!R^n \to I\!R$, let $\bar{x}$ be a local minimum of f on C and let f be differentiable at $\bar{x}$. Then there exist numbers $y_i \geq 0,\; i \in I(\bar{x})$ such that

$$\nabla f(\bar{x}) + \sum_{i \in I(\bar{x})} y_i b_i = 0. \tag{6}$$

If in addition f is convex on C, then condition (6) is necessary and sufficient for a global minimum at $\bar{x}$. Note that if we set $y_i = 0$ for $i \notin I(\bar{x})$, then condition (6) is written as

$$f(\bar{x}) + B^{\mathsf{T}} y = 0,$$

where the vector $y \in I\!R^m$ has components y_i and B is the matrix with rows b_i, $i = 1, \ldots, m$. The numbers y_i are called *Lagrange multipliers* associated with the solution $\bar{x}$.

Specifically, let $\bar{x}$ be a local solution of the problem

$$\min f(x) \quad \text{subject to} \quad Bx \leq a, \tag{7}$$

where $x \in I\!R^n$, B is an $m \times n$ matrix, and $a = (a_1, \ldots, a_m) \in I\!R^m$. Then, there exists a vector $\bar{y} \in I\!R^m$ of Lagrange multipliers such that $(\bar{x}, \bar{y})$ is a solution of the following *optimality system*:

$$\nabla f(x) + B^T y = 0, \quad Bx \leq a, \quad y \geq 0, \quad \langle Bx - a, y \rangle = 0. \tag{8}$$

If in addition f is convex on $I\!R^n$, then the optimality system (8) is also a sufficient condition for a global minimum.

Note that $\langle Bx - a, y \rangle = 0$ means that if $(Bx - a)_i = 0$, then $y_i \geq 0$, and if $(Bx - a)_i < 0$, then $y_i = 0$. This condition is sometimes called the *complementarity* condition .

A function $L : I\!R^n \times I\!R^m \to I\!R$ of the form

$$L(x, y) = f(x) + \langle y, Bx - a \rangle$$

is said to be a *Lagrange function* for problem (7). Note that the first relation (8) is just the Fermat rule "derivative equals zero" for L with respect to x. Let $I\!R^m_+$ be the positive orthant in $I\!R^m$ and let $E = I\!R^n \times I\!R^m_+$. Then, in terms of the Lagrange function, the optimality system (8) can be written as a variational inequality of the form

$$\begin{pmatrix} \nabla_x L(x, y) \\ -\nabla_y L(x, y) \end{pmatrix} + N_E(x, y) \ni (0, 0).$$

Note that we define the Lagrange function L by multiplying the constraints by a "multiplier" and then minimize L with respect to x without constraints to obtain a solution of the problem. This "trick" is sometimes called the *Lagrange principle* in the literature. Of course, the main difficulty here is to find Lagrange multipliers.

The idea to use a Lagrange function can be extended to optimization problems that are much more general than (7). As an example, which we will deal with later in the book, consider the problem

$$\min f(x, u) \quad \text{subject to} \quad g(x, u) = 0, \quad u \in C, \tag{9}$$

where $f : I\!R^n \times I\!R^m \to I\!R$ and $g : I\!R^n \times I\!R^m \to I\!R^k$ are continuously differentiable functions, and C is a convex and closed set in $I\!R^m$. Here the minimization is with respect to both x and u. We introduce here a Lagrange function involving a multiplier vector $y \in I\!R^k$ associated with the equality constraints, of the form

$$L(x, u, y) = f(x, u) + y^{\mathsf{T}} g(x, u).$$

Then we apply the Lagrange principle by minimizing L with respect to (x, u) without the equality constraint, that is, minimizing $L(x, u)$ subject to $(x, u) \in \mathbb{R}^n \times C$. The resulting first-order optimality condition then becomes

$$\begin{aligned} &\nabla_x L(x, u, y) = 0, \\ &\nabla_u L(x, u, y) + N_C(u) \ni 0. \end{aligned} \tag{10}$$

There is an important issue here; namely, in contrast to problem (7) this optimality condition is valid on an additional assumption about the constraints, the so-called *constraint qualification condition.* Without going into detail, we only mention that when $C = \mathbb{R}^m$, the following constraint qualification condition implies the validity of the optimality condition (10): if $(\bar{x}, \bar{u})$ is a solution of problem (9) then rank $\nabla_x g(\bar{x}, \bar{u}) = k$.

Let C be a polyhedral set as in (5). Recall that the *tangent cone* to C at x has the form

$$T_C(x) = \{ w \,|\, \langle b_i, w \rangle \le 0 \ \text{ for } \ i \in I(x) \},$$

while the normal cone to C at x is

$$N_C(x) = \Big\{ v \,\Big|\, v = \sum_{i=1}^{m} y_i b_i \ \text{ with } y_i \ge 0 \text{ for } i \in I(x),\ y_i = 0 \text{ for } i \notin I(x) \Big\}.$$

Critical Cone. *For $x \in C$ and $v \in N_C(x)$, the critical cone to C at x for v is defined as*

$$K_C(x, v) = \{ w \in T_C(x) \,|\, w \perp v \},$$

where $w \perp v$ means that w is orthogonal to v, that is, $\langle w, v \rangle = 0$ (Fig. 1.2).

Example. The nonnegative orthant $\mathbb{R}^n_+$ is a polyhedral cone in $\mathbb{R}^n$, since it consists of the vectors $x = (x_1, \ldots, x_n)$ satisfying the linear inequalities $x_j \ge 0$, $j = 1, \ldots, n$. For $v = (v_1, \ldots, v_n)$, one has

$$v \in N_{\mathbb{R}^n_+}(x) \iff x_j \ge 0,\ v_j \le 0,\ x_j v_j = 0 \ \text{ for } \ j = 1, \ldots, n.$$

Thus, whenever $v \in N_{\mathbb{R}^n_+}(x)$, one has, in terms of the index sets

$$\begin{aligned} J_1 &= \{ j \,|\, x_j > 0,\ v_j = 0 \}, \\ J_2 &= \{ j \,|\, x_j = 0,\ v_j = 0 \}, \\ J_3 &= \{ j \,|\, x_j = 0,\ v_j < 0 \}, \end{aligned}$$

that the vectors $w = (w_1, \ldots, w_n)$ belonging to the critical cone to $\mathbb{R}^n_+$ at x for v are characterized by

$$w \in K_{\mathbb{R}^n_+}(x, v) \iff \begin{cases} w_j \ \text{free} & \text{for } j \in J_1, \\ w_j \ge 0 & \text{for } j \in J_2, \\ w_j = 0 & \text{for } j \in J_3. \end{cases}$$

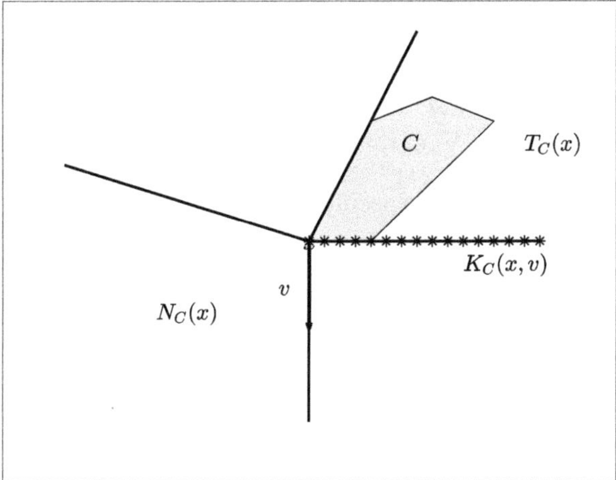

Fig. 1.2: Tangent, normal, and critical cones to a polyhedral set

The critical cones are involved in the expression of second-order optimality conditions.

Second-Order Optimality Conditions. *Let C be a polyhedral convex set in $I\!R^n$ and let $g : I\!R^n \to I\!R$ be twice continuously differentiable on an open set containing C. Let $\bar{x} \in C$ and $\bar{v} = -\nabla g(\bar{x})$.*

(a) (Necessary condition) *If g has a local minimum on C at $\bar{x}$, then $\bar{x}$ satisfies the first-order necessary optimality condition stated above and has $\langle w, \nabla^2 g(\bar{x}) w\rangle \geq 0$ for all $w \in K_C(\bar{x}, \bar{v})$.*

(b) (Sufficient condition) *If $\bar{x}$ satisfies the first-order necessary optimality condition stated above and also $\langle w, \nabla^2 g(\bar{x}) w\rangle > 0$ for all nonzero $w \in K_C(\bar{x}, \bar{v})$, then g has a local minimum on C at $\bar{x}$, indeed a strong local minimum in the sense of there being an $\varepsilon > 0$ such that*

$$g_0(x) \geq g_0(\bar{x}) + \frac{\varepsilon}{2}|x - \bar{x}|^2 \quad \textit{for all } x \textit{ near } \bar{x} \textit{ satisfying the constraints.}$$

Nonlinear Programming. We consider next the basic problem of nonlinear programming, of the form:

$$\text{minimize } g_0(x) \text{ over all } x \text{ satisfying } g_i(x) \begin{cases} \leq 0 & \text{for } i \in [1, s], \\ = 0 & \text{for } i \in [s+1, m], \end{cases} \tag{11}$$

where $g_i : I\!R^n \to I\!R$, $i = 0, 1, \ldots, m$ and $1 \leq s \leq m$. In terms of the Lagrange function

$$L(x, y) = g_0(x) + y_1 g_1(x) + \cdots + y_m g_m(x),$$

where $y = (y_1, \ldots, y_m)$ is the vector of the Lagrange multipliers, the variational inequality representing the associated first-order necessary condition is of the form

$$\begin{pmatrix} \nabla_x L(x, y) \\ -\nabla_y L(x, y) \end{pmatrix} + N_E(x, y) \ni (0, 0),$$

where $E = I\!R^n \times [I\!R^s_+ \times I\!R^{m-s}]$. Specifically, the optimality system for problem (11) has the form

$$\begin{aligned} &y \in I\!R^s_+ \times I\!R^{m-s}, \quad g_i(x) \begin{cases} \le 0 & \text{for } i \in [1, s] \text{ with } y_i = 0, \\ = 0 & \text{for all other } i \in [1, m], \end{cases} \\ &\nabla g_0(x) + y_1 \nabla g_1(x) + \cdots + y_m \nabla g_m(x) = 0. \end{aligned} \tag{12}$$

The system (12) is called the *Karush–Kuhn–Tucker condition* for the nonlinear programming problem (11). The existence of y satisfying these conditions with $x = \bar{x}$ is necessary for the local optimality of $\bar{x}$ under a *constraint qualification condition*. We adopt here the *Mangasarian–Fromovitz* constraint qualification condition, which has the following form:

$$\begin{aligned} &\exists\, w \in I\!R^n \quad \text{with} \quad \begin{cases} \nabla g_i(\bar{x}) w < 0 & \text{for } i \in [1, s] \text{ with } g_i(\bar{x}) = 0, \\ \nabla g_i(\bar{x}) w = 0 & \text{for } i \in [s+1, m], \end{cases} \\ &\text{and the vectors } \nabla g_i(\bar{x}) \text{ for } i \in [s+1, m] \text{ are linearly independent.} \end{aligned}$$

The Mangasarian–Fromovitz condition is implied by the condition of *linear independence of the gradients of the active constraints*: the vectors $\nabla g_i(\bar{x})$ for $i \in [1, s]$ with $g_i(\bar{x}) = 0$ and $i \in [s+1, m]$ are linearly independent.

Under the Mangasarian–Fromovitz condition, the existence of y satisfying (12) is not only necessary but also sufficient for the global optimality of x as long as $L(x, y)$ is convex as a function of $x \in I\!R^n$ for each fixed $y \in I\!R^s_+ \times I\!R^{m-s}$, which is equivalent to having

$$g_0, g_1, \ldots, g_s \ \text{ convex, and } \ g_{s+1}, \ldots, g_m \ \text{ affine.}$$

Then (11) is a problem of *convex programming*.

We state next necessary second-order conditions as well as sufficient second-order conditions for local optimality for the nonlinear programming problems (11):

Second-Order Optimality in Nonlinear Programming. *Let $\bar{x}$ be a point satisfying the constraints in* (11). *Let $I(\bar{x})$ be the set of indices i of the active constraints at $\bar{x}$, and suppose that the gradients $\nabla g_i(\bar{x})$ for $i \in I(\bar{x})$ are linearly independent. Let K consist of the vectors $w \in I\!R^n$ satisfying*

$$\langle \nabla g_i(\bar{x}), w \rangle \begin{cases} \le 0 & \textit{for } i \in I(\bar{x}) \textit{ with } i \le s, \\ = 0 & \textit{for all other } i \in I(\bar{x}) \textit{ and also for } i = 0. \end{cases}$$

(a) (Necessary condition) *If $\bar{x}$ furnishes a local minimum in problem* (11), *then a multiplier vector $\bar{y}$ exists such that $(\bar{x}, \bar{y})$ not only satisfies* (12) *but also has*

$$\langle w, \nabla^2_{xx} L(\bar{x}, \bar{y}) w \rangle \geq 0 \textit{ for all } w \in K. \tag{13}$$

(b) (Sufficient condition) *If a multiplier vector* $\bar{y}$ *exists such that* $(\bar{x}, \bar{y})$ *satisfies the conditions in* (12) *and if* (13) *holds with strict inequality when* $w \neq 0$, *then* $\bar{x}$ *furnishes a strong local minimum for problem* (11).

Lecture 2
Continuity of Set-Valued Mappings

Recall that a function f acting from a metric space (X, ρ) to a metric space (Y, ρ) is continuous at a point $\bar{x} \in \text{dom } f$ if for every $\varepsilon > 0$ there exists $\delta > 0$ such that $\rho(f(x), f(\bar{x})) \leq \varepsilon$ whenever $\rho(x, \bar{x}) \leq \delta$. A function f is said to be continuous on, or over, or relative to, a set $D \subset \text{dom } f$ if it is continuous at every $x \in D$.

Defining a notion of continuity of a set-valued mapping is a much more challenging task. Indeed, there are a number of definitions for convergence of sequences of sets and also various ways to define when two sets are "close" to each other. In this book we will use the following definitions:

Outer Semicontinuity. *A set-valued mapping $F : X \rightrightarrows Y$ is said to be outer semicontinuous at $\bar{x} \in$ dom F if for every open set V such that $V \supset F(\bar{x})$ there exists a neighborhood U of $\bar{x}$ such that $F(x) \subset V$ for every $x \in U$.*

Clearly, the definition of outer semicontinuity of a set-valued mapping is a generalization of the epsilon-delta definition given above for functions. However, in order to define continuity of set-valued mappings we need in addition the following property:

Inner Semicontinuity. *A set-valued mapping $F : X \rightrightarrows Y$ is said to be inner semicontinuous at $\bar{x} \in$ dom F if for every open set V such that $F(\bar{x}) \cap V \neq \emptyset$ there exists a neighborhood U of $\bar{x}$ such that $F(x) \cap V \neq \emptyset$ for every $x \in U$.*

Putting the outer and inner semicontinuity together, a set-valued mapping F is said to be *continuous* at $\bar{x}$ if it is both outer and inner semicontinuous at $\bar{x}$.

Sometimes it is more convenient to use the following equivalent definition of inner semicontinuity: for every sequence $x^k \to \bar{x}$ and every $\bar{y} \in F(\bar{x})$ there exists a sequence $y^k \in F(x^k)$ such that $y^k \to \bar{y}$. Outer semicontinuity at $\bar{x}$ in turn implies, but is not equivalent to, the property that for every sequence $x_k \to \bar{x}$, any cluster point of a sequence $y_k \in F(x_k)$ is an element of $F(\bar{x})$. This implies in particular that the value $F(\bar{x})$ of a mapping F, which is outer semicontinuous at $\bar{x}$, is closed.

© The Author(s), under exclusive license to Springer Nature Switzerland AG 2021

A. L. Dontchev, *Lectures on Variational Analysis*, Applied Mathematical Sciences 205, https://doi.org/10.1007/978-3-030-79911-3_2

If a mapping F is outer semicontinuous at every point in a set $D \subset \operatorname{dom} F$, then it is said to be outer semicontinuous on, or over, or relative to D, and the same for inner semicontinuity and continuity. In particular, any mapping that is outer semicontinuous in its domain has closed graph. In the literature, outer semicontinuity is often called upper semicontinuity, while inner semicontinuity is called lower semicontinuity. We reserve these names for real-valued functions, as defined in the preparatory section.

Examples. The mapping

$$F(x) = \begin{cases} [-1, 1] & \text{for } x < 0, \\ \{0\} & \text{for } x \geq 0 \end{cases}$$

is inner semicontinuous at 0 but not outer semicontinuous there. Indeed, the only element 0 of $F(0)$ has neighborhoods that are much smaller than $F(x)$ for $x < 0$. On the other hand, $0 = F(0)$ can be approached by the sequence of zeroes, the elements of which are in $F(x)$ for both $x < 0$ and $x > 0$. Similarly, the mapping

$$F(x) = \begin{cases} [-1, 1] & \text{for } x = 0, \\ \{0\} & \text{for } x \neq 0 \end{cases}$$

is outer semicontinuous at 0 but not inner semicontinuous there.

As an application, we consider continuity properties of solutions to the following optimization problem: for a parameter p that ranges over a metric space P, a function $g : P \times X \to I\!R$, and a mapping $D : P \rightrightarrows X$,

$$\text{minimize } g(p, x) \text{ subject to } x \in D(p).$$

Here, for any $p \in P$, $g(p, \cdot)$ is the objective function to be minimized with respect to x and $D(p)$ is the feasible set. As before, X is a metric space.

We consider the following two mappings: the *optimal value* mapping defined as

$$S_{\text{val}} : P \ni p \mapsto \inf_x \left\{ g(p, x) \,\middle|\, x \in D(p) \right\},$$

and the *optimal solution* mapping defined as

$$S_{\text{opt}} : P \ni p \mapsto \left\{ x \in D(p) \,\middle|\, g(p, x) = S_{\text{val}}(p) \right\}.$$

Recall that a set D in a metric space is compact when every sequence $\{x_n\}$ with $x_n \in D$ has a subsequence that is convergent to an element of D.

Theorem 2.1. *Let $\bar{p} \in P$ be fixed and let the feasible set $D(\bar{p})$ be nonempty and compact. Suppose that:*

(a) *the mapping $p \mapsto D(p)$ is continuous at $\bar{p}$;*

(b) *the function g is continuous at $(\bar{p}, \bar{x})$ for every $\bar{x} \in D(\bar{p})$.*

Then the optimal value mapping S_{val} is continuous at $\bar{p}$, whereas the optimal solution mapping S_{opt} is outer semicontinuous at $\bar{p}$.

Proof. By the Weierstrass theorem, $S_{\text{opt}}(\bar{p}) \neq \emptyset$. Let $\bar{x} \in S_{\text{opt}}(\bar{p})$. From the inner semicontinuity of D, for every sequence $p_k \to \bar{p}$, there exists a sequence of points $x_k \in D(p_k)$ such that $x_k \to \bar{x}$ as $k \to \infty$. Choose $p_k \to \bar{p}$ and a corresponding sequence $x_k \in D(p_k)$ such that $x_k \to \bar{x}$.

Let $\varepsilon > 0$. Then, from the continuity of g, for all k sufficiently large,

$$S_{\text{val}}(p_k) \leq g(p_k, x_k) \leq g(\bar{p}, \bar{x}) + \varepsilon = S_{\text{val}}(\bar{p}) + \varepsilon.$$

Since ε is arbitrary, this yields

$$\limsup_{p \to \bar{p}} S_{\text{val}}(p) \leq S_{\text{val}}(\bar{p}). \tag{1}$$

Hence, the value function is upper semicontinuous at $\bar{p}$.

The proof of the lower semicontinuity of S_{val} is a bit more complicated. Let $\varepsilon > 0$. From the continuity of g at $(\bar{p}, \bar{x})$, for every $x \in D(\bar{p})$, there exist neighborhoods $U(x)$ of x and $W(x)$ of $\bar{p}$, both depending on the choice of x, such that

$$g(\bar{p}, x) \leq g(p, x') + \varepsilon \quad \text{whenever } p \in W(x) \text{ and } x' \in U(x). \tag{2}$$

Since $D(\bar{p})$ is compact, from the open cover $\cup\{U(x) \mid x \in D(\bar{p})\}$ of $D(\bar{p})$, one can extract a finite subcover $\cup_{i=1}^m \{U(x_i) \mid x_i \in D(\bar{p}), i = 1, \ldots, m\}$, which clearly is an open set containing $D(\bar{p})$. The assumed outer semicontinuity of D at $\bar{p}$ implies that there exists a neighborhood V of $\bar{p}$ such that

$$D(p) \subset \cup_{i=1}^m U(x_i) \quad \text{whenever } p \in V. \tag{3}$$

As well known, the intersection $\cap_{i=1}^m W(x_i)$ of finitely many open sets $W(x_i)$ is an open set; hence, the set

$$S = V \cap \left(\cap_{i=1}^m W(x_i)\right)$$

is an open neighborhood of $\bar{p}$.

Fix any $p \in S$ and any $x \in D(p)$. From (2) we have $x \in U(x_j)$ for some j from 1 to m. Since $p \in W(x_j)$, from (2), we get

$$g(\bar{p}, x_j) \leq g(p, x) + \varepsilon.$$

This yields

$$S_{\text{val}}(\bar{p}) \leq g(p, x) + \varepsilon,$$

hence

$$S_{\text{val}}(\bar{p}) \leq S_{\text{val}}(p) + \varepsilon.$$

Since p is an arbitrarily chosen point in a neighborhood S of $\bar{p}$, this gives us lower semicontinuity of S_{val}.

We now show the outer semicontinuity of the mapping S_{opt} at $\bar{p}$. On the contrary, suppose that there exist $x \notin S_{\text{opt}}(\bar{p})$ and sequences $p_k \to \bar{p}$ and $x_k \to x$ as $k \to \infty$ such that $x_k \in S_{\text{opt}}(p^k)$.

Then $x_k \in D(p_k)$ and from the outer semicontinuity of D at $\bar{p}$ we have that $x \in D(\bar{p})$. But then, as we already proved,

$$g(p_k, x_k) = S_{\text{val}}(p_k) \longrightarrow S_{\text{val}}(\bar{p}) \quad \text{as } k \to \infty.$$

By continuity of g,

$$g(p_k, x_k) \longrightarrow g(\bar{p}, x),$$

hence

$$g(\bar{p}, x) = S_{\text{val}}(\bar{p}),$$

that is, $x \in S_{\text{opt}}(\bar{p})$, contradiction. The proof is complete. □

The required compactness of $D(\bar{p})$ in Theorem 2.1 cannot be dropped, in general. Indeed, take $X = P = Y = I\!R$, $g(p, x) = px$, $D(p) = X$ for all p, $\bar{p} = 0$. Then $S_{\text{val}}(p) = -\infty$ for $p \neq \bar{p}$ and $S_{\text{val}}(\bar{p}) = 0$. Clearly, the lack of compactness of $D(\bar{p})$ yields that the optimal values are unbounded from below and there is no solution for $\bar{p}$. If our goal is to prove existence, the compactness assumption can be avoided by imposing conditions on the objective function. One such condition is the coercivity condition we utilize later in the book. Basically, existence could be ensured provided that there exists a compact minimizing sequence; we leave the proof of the corresponding statement to the interested reader.

We introduce next the concept of Lipschitz continuity of set-valued mappings. For that we need to define the notion of distance between sets in a metric space X with metric ρ.

Excess. *For sets C and D in X, the excess of C beyond D is defined by*

$$e(C, D) = \sup_{x \in C} d(x, D),$$

with the convention that $e(\emptyset, D) = 0$ when $D \neq \emptyset$ and $+\infty$ otherwise.

Pompeiu–Hausdorff Distance. *Given sets C and D in a metric space X, the quantity*

$$h(C, D) = \max\{e(C, D), e(D, C)\}$$

is said to be the Pompeiu–Hausdorff distance between C and D (Fig. 2.1).

Note that when X is a linear space, then the excess and the Pompeiu–Hausdorff distance can be expressed as

$$e(C, D) = \inf\{\, \tau \geq 0 \,|\, C \subset D + \tau I\!B \}$$

and

$$h(C, D) = \inf\{\, \tau \geq 0 \,|\, C \subset D + \tau I\!B,\ D \subset C + \tau I\!B \}.$$

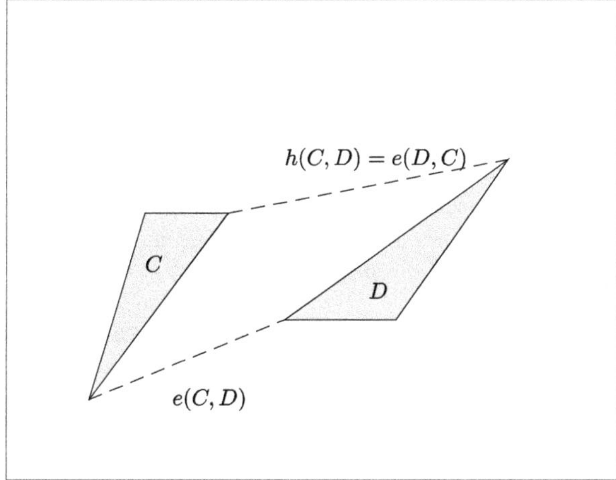

Fig. 2.1: Excess and Pompeiu–Hausdorff distance

The Pompeiu–Hausdorff distance is unaffected by whether C and D are closed or not. Note that both $e(C, D)$ and $h(C, D)$ could be $+\infty$ when unbounded sets are involved. For that reason in particular, the Pompeiu–Hausdorff distance does not furnish a metric on the space of nonempty closed subsets of X, although it does on the space of nonempty closed subsets of a bounded set $W \subset X$. Also note that $e(C, \emptyset) = +\infty$ for any set C, including the empty set.

The definition of continuity of a set-valued mapping F at $\bar{x}$ introduced in the beginning of this lecture is equivalent to having $\lim_{x\to\bar{x}} h(F(x), F(\bar{x})) = 0$. A quantitative notion of continuity of set-valued mappings can be introduced with the help of the Pompeiu–Hausdorff distance between sets, in the same way as Lipschitz continuity is defined for functions.

Lipschitz Continuity of Set-Valued Mappings. *Let X and Y be metric spaces. A mapping $F : X \rightrightarrows Y$ is said to be Lipschitz continuous relative to a (nonempty) set D in X if $D \subset \operatorname{dom} F$, the set $F(x)$ is closed for all $x \in D$, and there exists $\kappa \geq 0$ called Lipschitz constant, such that*

$$h(F(x'), F(x)) \leq \kappa\rho(x', x) \quad \textit{for all } x', x \in D.$$

When Y is a linear space, then this definition can be stated equivalently as follows: there exists $\kappa \geq 0$ such that

$$F(x') \subset F(x) + \kappa\rho(x', x)I\!B \quad \text{for all } x', x \in D. \tag{4}$$

Note that if F satisfies (4) and $F(x)$ is bounded for some $x \in D$, then it is bounded for all $x \in D$. When F is single-valued on D, we obtain from this definition the notion of Lipschitz continuity of a function. The requirement in the definition that F be closed-valued may seem redundant; indeed, it is added only to avoid the paradox that Lipschitz continuity does not imply continuity. Recall the definition of continuity according to which a mapping that is continuous necessarily does have closed values.

Theorem 2.2. *In the context of Theorem 2.1, suppose that for every $p \in P$ the feasible set $D(p)$ is nonempty and compact, the mapping $p \mapsto D(p)$ is Lipschitz continuous on P, and the function g is Lipschitz continuous on $P \times X$. Then the optimal value mapping S_{val} is Lipschitz continuous on P.*

Proof. Let κ be a Lipschitz constant for D and μ be a Lipschitz constant for g. Denote by ρ the metrics of both X and P. Choose any $p, p' \in P$, and any $x \in S_{\text{opt}}(p)$, which exists because $D(p)$ is nonempty and compact. In particular, we have $x \in D(p)$. From the Lipschitz continuity of D, there exists $x' \in D(p')$ such that $\rho(x', x) \leq \kappa\rho(p', p)$. Using this along with the fact that $S_{\text{val}}(p') \leq g(p', x')$ and $S_{\text{val}}(p) = g(p, x)$ we obtain

$$S_{\text{val}}(p') - S_{\text{val}}(p) \leq g(p', x') - g(p, x) \leq \mu(\rho(p', p) + \rho(x', x)) \leq \mu(1 + \kappa)\rho(p', p).$$

The symmetry of p and p' gives us the desired result. □

Lecture 3
Lipschitz Continuity of Polyhedral Mappings

In this lecture we will focus on a particular class of set-valued mappings in finite dimensions, with significant applications in optimization, which are automatically Lipschitz continuous. Recall that a set in $\mathbb{R}^n$ is *polyhedral* if it can be expressed as the intersection of a finite collection of closed half-spaces and/or hyperplanes. In other words, a set $D \subset \mathbb{R}^n$ is polyhedral when it can be represented as

$$D = \{x \in \mathbb{R}^n \mid Ax \leq d\} \tag{1}$$

for a natural number r, an $r \times n$ matrix A, and a vector $d \in \mathbb{R}^r$.

Polyhedral Mappings. *A mapping $S : \mathbb{R}^m \rightrightarrows \mathbb{R}^n$ is said to be polyhedral if its graph is a polyhedral set.*

Directly from the definition of a polyhedral set and (1), polyhedrality of a mapping $S : \mathbb{R}^m \rightrightarrows \mathbb{R}^n$ is equivalent to the existence of a positive integer r, matrices $D \in \mathbb{R}^{r\times n}$, $E \in \mathbb{R}^{r\times m}$, and a vector $q \in \mathbb{R}^r$ such that

$$S(y) = \left\{ x \in \mathbb{R}^n \,\middle|\, Dx + Ey \leq q \right\} \quad \text{for all } \; y \in \mathbb{R}^m. \tag{2}$$

For instance, any mapping S whose graph is a linear subspace is a polyhedral mapping.

The following theorem establishes Lipschitz continuity of polyhedral mappings.

Theorem 3.1. *Any polyhedral mapping $S : \mathbb{R}^m \rightrightarrows \mathbb{R}^n$ is Lipschitz continuous in its domain.*

We will prove this theorem by employing a fundamental result due to A. J. Hoffman regarding approximate solutions of systems of linear inequalities. For a vector $a = (a_1, a_2, \ldots, a_n) \in \mathbb{R}^n$, we use the vector notation that

$$a_+ = (\max\{0, a_1\}, \ldots, \max\{0, a_n\}).$$

© The Author(s), under exclusive license to Springer Nature Switzerland AG 2021

A. L. Dontchev, *Lectures on Variational Analysis*, Applied Mathematical Sciences 205, https://doi.org/10.1007/978-3-030-79911-3_3

Lemma 3.2. (Hoffman) *Let A be a nonzero $m \times n$ matrix. Consider the set-valued mapping*

$$S : y \mapsto \{\xi \in I\!R^n \,|\, A\xi \leq y\} \quad \text{for } y \in I\!R^m.$$

Then there exists a constant L such that for every $y \in \operatorname{dom} S$ and every $x \in I\!R^n$ one has

$$d(x, S(y)) \leq L\|(Ax - y)_+\|. \tag{3}$$

Before going further, note that the mapping S in Hoffman lemma can be written as

$$S(y) = \{x \in I\!R^n \mid y - Ax \in I\!R^m_+\}.$$

Let $F(x) = Ax + I\!R^m_+$. Then $S(y) = F^{-1}(y)$ and

$$d(y, F(x)) = d(y, Ax + I\!R^m_+) = \|(Ax - y)_+\|.$$

In these terms, the Hoffman lemma says that

$$d(x, F^{-1}(y)) \leq Ld(y, F(x)).$$

This is the same as saying that the mapping $F = A + I\!R^m_+$ is *metrically regular globally*. We will define metric regularity in the following lecture.

Proof of Theorem 3.1. Let $y, y' \in \operatorname{dom} S$, and let $x \in S(y)$. Since S is polyhedral, from the representation (2), we have $Dx + Ey - q \leq 0$. Observe that

$$Dx + Ey' - q = Dx + Ey - q - Ey + Ey' \leq -Ey + Ey'. \tag{4}$$

From Lemma 3.2 we obtain the existence of a constant L such that

$$d(x, S(y')) \leq L\|(Dx + Ey' - q)_+\|,$$

hence, by (4),

$$d(x, S(y')) \leq L\|(E(y' - y))_+\| \leq L\|E(y - y')\|.$$

Since x is arbitrarily chosen in $S(y)$, this leads to

$$e(S(y), S(y')) \leq \kappa\|y - y'\|$$

with $\kappa = L\|E\|$. The same must hold with the roles of y and y' reversed, and in consequence S is Lipschitz continuous on $\operatorname{dom} S$. □

Application to Linear Programming. Consider the following problem of linear programming in which y acts as a parameter:

$$\text{minimize } \langle c, x\rangle \text{ over all } x \in I\!R^n \text{ satisfying } Ax \leq y. \tag{5}$$

Here c is a fixed vector in $I\!R^n$, and A is a fixed matrix in $I\!R^{m\times n}$. Define the following mappings associated with (5): the feasible set mapping

$$S_{\text{feas}} : y \mapsto \{ x \,|\, Ax \le y \}, \tag{6}$$

the optimal value mapping

$$S_{\text{val}} : y \mapsto \inf_x \{ \langle c, x\rangle \,|\, Ax \le y \} \text{ when the inf is finite,} \tag{7}$$

and the optimal set mapping by

$$S_{\text{opt}} : y \mapsto \{ x \in S_{\text{feas}}(y) \,|\, \langle c, x\rangle = S_{\text{val}}(y) \}. \tag{8}$$

It is known from the theory of linear programming that $S_{\text{opt}}(y) \neq \emptyset$ when the infimum in (7) is finite (and only then). The following theorem establishes Lipschitz continuity of solution mappings in linear programming.

Theorem 3.3. *The mappings in* (6), (7), *and* (8) *are Lipschitz continuous relative to their domains, the domain in the case of* (7) *and* (8) *being the set D consisting of all y for which the infimum in* (7) *is finite.*

Proof. The Lipschitz continuity of S_{feas} comes from Theorem 3.1. Then from Theorem 2.2 we obtain that the optimal value S_{val} is Lipschitz continuous on its domain D. For the case of S_{opt}, consider the set-valued mapping

$$I\!R^m \times I\!R \ni (y, t) \mapsto G(y, t) = \{ x \in I\!R^n \,|\, Ax \le y,\ \langle c, x\rangle \le t \}.$$

Since this mapping is polyhedral we can apply Theorem 3.1. The proof is completed by observing that $S_{\text{opt}}(y) = G(y, S_{\text{val}}(y))$ for $y \in D$ and using the Lipschitz continuity of S_{val}. □

In the reminder of this lecture we consider a "one-point" property of set-valued mappings, by fixing one of the points y and y' in the definition of Lipschitz continuity at its reference value $\bar{y}$. Then these points no longer play symmetric roles, so we use the excess instead of the Pompeiu–Hausdorff distance.

Outer Lipschitz Continuity. *A mapping $S : I\!R^m \rightrightarrows I\!R^n$ is said to be outer Lipschitz continuous at $\bar{y}$ relative to a set D if $\bar{y} \in D \subset \operatorname{dom} S$, $S(\bar{y})$ is a closed set, and there is a constant $\kappa \ge 0$ along with a neighborhood V of $\bar{y}$ such that*

$$e(S(y), S(\bar{y})) \le \kappa |y - \bar{y}| \quad \textit{for all} \;\; y \in V \cap D,$$

or equivalently

$$S(y) \subset S(\bar{y}) + \kappa |y - \bar{y}| I\!B \quad \textit{for all} \;\; y \in V \cap D.$$

If S is outer Lipschitz continuous at every point $y \in D$ relative to D with the same κ, then S is said to be outer Lipschitz continuous relative to D.

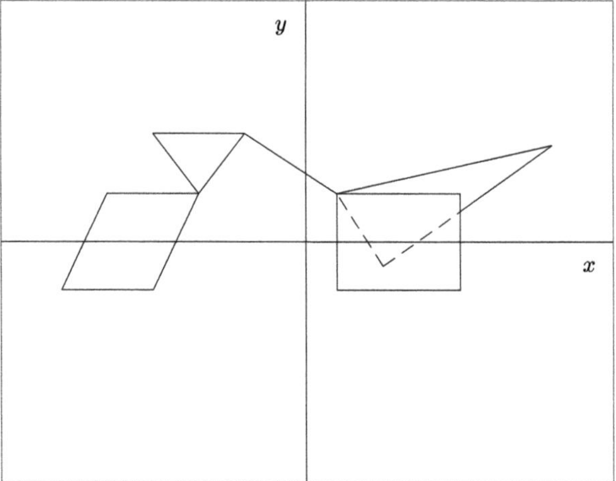

Fig. 3.1: Graph of a piecewise polyhedral map

It is clear that any mapping that is Lipschitz continuous relative to a set D with constant κ is also outer Lipschitz continuous relative to D with constant κ. Also, outer Lipschitz continuity at a point $\bar{y}$ implies outer semicontinuity at $\bar{y}$. Then from the discussion in the beginning of Lecture 2 we obtain that the values of outer Lipschitz continuous mappings are closed sets. For single-valued mappings, outer Lipschitz continuity becomes calmness, a property introduced in the preparatory section. Outer Lipschitz continuity is also called upper Lipschitz continuity.

We present next a result that historically was the main motivation for introducing the property of outer Lipschitz continuity. It uses the following concept.

Piecewise Polyhedral Mapping. *A set-valued mapping $S : I\!R^n \rightrightarrows I\!R^m$ is said to be piecewise polyhedral if* gph *S is the union of finitely many sets that are polyhedral in $I\!R^n \times I\!R^m$.*

In other words, the graph of a piecewise polyhedral mapping is comprised by finitely many polyhedral "pieces" that could overlap. An illustration of the graph of a piecewise polyhedral mapping is illustrated in Fig. 3.1. Any polyhedral mapping as defined earlier in this lecture is obviously piecewise polyhedral; its graph is comprised of only one "piece." Clearly, a piecewise polyhedral mapping has closed graph, since polyhedral convex sets are closed. Then, such a mapping is outer semicontinuous in its domain. But more could be said about continuity properties of piecewise polyhedral mappings, as stated in the following theorem proved at the end of the lecture.

Theorem 3.4. *Any piecewise polyhedral mapping $S : I\!R^m \rightrightarrows I\!R^n$ is outer Lipschitz continuous in its domain.*

The following exercise provides an important example of a piecewise polyhedral mapping.

Exercise 3.5. *Given an $n \times n$ matrix A and a polyhedral set C in $\mathbb{R}^n$, show that the solution mapping of the linear variational inequality*

$$v \mapsto S(v) = \{x \mid v \in Ax + N_C(x)\} \quad \text{for } v \in \mathbb{R}^n$$

is outer Lipschitz continuous relative to its domain.

Guide. Any polyhedral set C can be represented (in a non-unique manner) by a system of affine inequalities:

$$C = \{x \mid \langle a_i, x\rangle \le \alpha_i \quad \text{for } i = 1, 2, \ldots, m\}.$$

Recall that the normal cone to C at the point $x \in C$ is the set

$$N_C(x) = \left\{u \mid u = \sum_{i=1}^m y_i a_i,\ y_i \ge 0 \text{ for } i \in I(x),\ y_i = 0 \text{ for } i \notin I(x)\right\},$$

where $I(x) = \{i \mid \langle a_i, x\rangle = \alpha_i\}$ is the active index set for $x \in C$. The graph of the normal cone mapping N_C is the union, with respect to all possible subsets J of $\{1, \ldots, m\}$, of the polyhedral sets

$$\left\{(x, u) \mid u = \sum_{i=1}^m y_i a_i,\ \langle a_i, x\rangle = \alpha_i,\ y_i \ge 0 \text{ if } i \in J,\ \langle a_i, x\rangle < \alpha_i,\ y_i = 0 \text{ if } i \notin J\right\}.$$

Proofs

Proof of Lemma 3.2. For any $y \in \operatorname{dom} S$ the set $S(y)$ is nonempty, convex, and closed; hence any point $x \notin S(y)$ has a unique (Euclidean) projection $u = P_{S(y)}(x)$ on $S(y)$ given by

$$u \in S(y),\ \|u - x\| = d(x, S(y)), \tag{9}$$

which is the solution of the problem

$$\min \frac{1}{2}\|u - x\|^2 \quad \text{subject to } u \in S(y).$$

We employ next the necessary optimality condition for this problem. Let $N_{S(y)}$ be the normal cone mapping to the convex set $S(y)$. Then, because of convexity, the problem of projecting x on $S(y)$ is equivalent to that of finding the unique u such that

$$0 \in u - x + N_{S(y)}(u).$$

The normal cone to a polyhedral set has the form

$$N_{S(y)}(u) = \left\{v \mid v = \sum_{i=1}^m \lambda_i a_i \text{ with } \lambda_i \ge 0,\ \lambda_i(\langle a_i, u\rangle - y_i) = 0,\ i = 1, \ldots, m\right\},$$

where the a_i's are the rows of the matrix A regarded as vectors in $\mathbb{R}^n$. Thus, the projection u of x on $S(y)$, as described by (9), can be obtained by finding a pair (u, λ) such that

$$\begin{cases} x - u - \sum_{i=1}^m \lambda_i a_i = 0, \\ \lambda_i \geq 0, \; \lambda_i(\langle a_i, u\rangle - y_i) = 0, \quad i = 1, \dots, m. \end{cases} \tag{10}$$

While the projection u exists and is unique, this variational inequality might not have a unique solution (u, λ) because the λ component might not be unique. But since $u \neq x$ (through our assumption that $x \notin S(y)$), we can conclude from the first relation in (10) that for any solution (u, λ) the vector $\lambda = (\lambda_1, \dots, \lambda_m)$ is not the zero vector. Consider the family $\mathcal{J}$ of subsets J of $\{1, \dots, m\}$ for which there are real numbers $\lambda_1, \dots, \lambda_m$ with $\lambda_i > 0$ for $i \in J$ and $\lambda_i = 0$ for $i \notin J$ and such that (u, λ) satisfies (10). Of course, if $\langle a_i, u\rangle - y_i < 0$ for some i, then $\lambda_i = 0$ according to the second relation (complementarity) in (10), and then this i cannot be an element of any J. That is,

$$J \in \mathcal{J} \text{ and } i \in J \quad \Longrightarrow \quad \langle a_i, u\rangle = y_i \text{ and } \lambda_i > 0. \tag{11}$$

Since the set of vectors λ such that (u, λ) solves (10) does not contain the zero vector, we have $\mathcal{J} \neq \emptyset$.

We will now prove that there is a nonempty index set $\bar{J} \in \mathcal{J}$ for which there are no numbers β_i, $i \in \bar{J}$ satisfying

$$\beta_i \geq 0, \; i \in \bar{J}, \quad \sum_{i \in \bar{J}} \beta_i > 0 \text{ and } \sum_{i \in \bar{J}} \beta_i a_i = 0. \tag{12}$$

On the contrary, suppose that for every $J \in \mathcal{J}$ this is not the case, that is, (12) holds with $\bar{J} = J$ for some β_i, $i \in J$. Let J' be a set in $\mathcal{J}$ with a minimal number of elements (J' might be not unique). Note that the number of elements in any J' is greater than 1. Indeed, if there were just one element i' in J', then we would have $\beta_{i'} a_{i'} = 0$ and $\beta_{i'} > 0$, hence $a_{i'} = 0$, and then, since (10) holds for (u, λ) such that $\lambda_i = \beta_i$ for $i = i'$, $\lambda_i = 0$ for $i \neq i'$, from the first equality in (10), we would get $x = u$, which contradicts the assumption that $x \notin S(y)$. Since $J' \in \mathcal{J}$, there are $\lambda'_i > 0$, $i \in J'$ such that

$$x - u = \sum_{i \in J'} \lambda'_i a_i. \tag{13}$$

By assumption, there are also real numbers $\beta'_i \geq 0$, $i \in J'$, such that

$$\sum_{i \in J'} \beta'_i > 0 \text{ and } \sum_{i \in J'} \beta'_i a_i = 0. \tag{14}$$

Multiplying both sides of the equality in (14) by a positive scalar t and subtracting from (13), we obtain

$$x - u = \sum_{i \in J'} (\lambda'_i - t\beta'_i) a_i.$$

Let

$$t_0 = \min_i \left\{ \frac{\lambda'_i}{\beta'_i} \;\middle|\; i \in J' \text{ with } \beta'_i > 0 \right\}.$$

Then for any $k \in J'$ for which this minimum is attained, we have

$$\lambda_i' - t_0\beta_i' \geq 0 \quad \text{for every } i \in J' \setminus k \quad \text{and} \quad x - u = \sum_{i \in J' \setminus k} (\lambda_i' - t\beta_i')a_i.$$

Thus, the vector $\lambda \in I\!R^m$ with components $\lambda_i' - t_0\beta_i'$ when $i \in J'$, and $\lambda_i' = 0$ when $i \notin J'$, is such that (u, λ) satisfies (10). Hence, we found a nonempty index set $J'' \in \mathcal{J}$ having fewer elements than J', which contradicts the choice of J'. The contradiction obtained proves that there is a nonempty index set $\bar{J} \in \mathcal{J}$ for which there are no numbers β_i, $i \in J$, satisfying (12). In particular, the zero vector in $I\!R^n$ is not in the convex hull $\operatorname{co}\{a_j, j \in \bar{J}\}$. Denote by $\bar{\lambda}_i > 0$ the multipliers corresponding to $i \in \bar{J}$. Note that $\sum_{j \in \bar{J}} \bar{\lambda}_i a_i \neq 0$, because otherwise (12) would hold for $\beta_i = \bar{\lambda}_i$. Then we have

$$\gamma = \sum_{i \in \bar{J}} \bar{\lambda}_i > 0.$$

Because (10) holds with $(u, \bar{\lambda})$, using (10) and (11), we obtain

$$\begin{aligned} d(0, \operatorname{co}\{a_j, j \in \bar{J}\})\|x - u\| &\leq \Big|\sum_{i \in \bar{J}} \frac{\bar{\lambda}_i}{\gamma} a_i\Big| \|x - u\| = \frac{1}{\gamma}\|x - u\|\|x - u\| \\ &= \Big\langle \frac{1}{\gamma}(x - u), x - u \Big\rangle = \Big\langle \frac{1}{\gamma}\Big(\sum_{i \in \bar{J}} \bar{\lambda}_i a_i\Big), x - u \Big\rangle \\ &= \sum_{i \in \bar{J}} \frac{\bar{\lambda}_i}{\gamma}(\langle a_i, x\rangle - \langle a_i, u\rangle) = \sum_{i \in \bar{J}} \frac{\bar{\lambda}_i}{\gamma}(\langle a_i, x\rangle - y_i) \\ &\leq \max_{i \in \bar{J}}\{(\langle a_i, x\rangle - y_i)_+\}. \end{aligned}$$

Hence, for some constant c independent of x and y, we have

$$d(x, S(y)) = \|x - u\| \leq c \max_{1 \leq i \leq m} \{(\langle a_i, x\rangle - y_i)_+\}.$$

This inequality remains valid (perhaps with a different constant c) after passing from the max vector norm to the equivalent Euclidean norm. This proves (3). □

Proof of Theorem 3.4. Let $\operatorname{gph} S = \bigcup_{i=1}^k G_i$ where the G_i's are polyhedral sets in $I\!R^m \times I\!R^n$. For each i define the mapping

$$S_i : y \mapsto \{x \,|\, (y, x) \in G_i\} \quad \text{for } y \in I\!R^m.$$

According to Theorem 3.1, each S_i is Lipschitz continuous on its domain. Let $\bar{y} \in \operatorname{dom} S$ and let

$$\mathcal{J} = \{i \,|\, \text{there exists } x \in I\!R^n \text{ with } (\bar{y}, x) \in G_i\}.$$

Then $\bar{y} \in \operatorname{dom} S_i$ for each $i \in \mathcal{J}$, and moreover,

$$S(\bar{y}) = \bigcup_{i \in \mathcal{J}} S_i(\bar{y}). \tag{15}$$

For any $i \notin \mathcal{J}$, since the sets $\{\bar{y}\} \times I\!R^n$ and G_i are disjoint and polyhedral, there is a neighborhood V_i of $\bar{y}$ such that $(V_i \times I\!R^n) \bigcap G_i = \emptyset$. Let $V = \bigcap_{i \notin \mathcal{J}} V_i$. Then of course V is a neighborhood of $\bar{y}$ and we have

$$(V \times I\!R^n) \bigcap \text{gph } S \subset \bigcup_{i=1}^{k} G_i \setminus \bigcup_{i \notin \mathcal{J}} G_i \subset \bigcup_{i \in \mathcal{J}} G_i. \tag{16}$$

Let $y \in V$. If $S(y) = \emptyset$; then there is nothing to prove. Let x be any point in $S(y)$. Then, from (16),

$$(y, x) \in (V \times I\!R^n) \bigcap \text{gph } S \subset \bigcup_{j \in \mathcal{J}} G_i;$$

hence for some $i \in \mathcal{J}$ we have $(y, x) \in G_i$, that is, $x \in S_i(y)$. Since each S_i is Lipschitz continuous and $\bar{y} \in \text{dom } S_i$, with constant κ_i, say, we obtain by using (15) that

$$d(x, S(\bar{y})) \leq \max_i d(x, S_i(\bar{y})) \leq \max_i e(S_i(y), S_i(\bar{y})) \leq \max_i \kappa_i \|y - \bar{y}\|.$$

Since x is an arbitrary point in $S(y)$, we conclude that S is outer Lipschitz continuous at $\bar{y}$ with constant $\kappa = \max_i \kappa_i$. □

Lecture 4
Metric Regularity

In this lecture we introduce and study metric regularity of set-valued mappings, a property that plays a central role in variational analysis. In what follows X and Y are metric spaces with metrics that are denoted in the same way by $\rho(\cdot,\cdot)$ but may be different. Recall that a set C in a metric space is *locally closed* at a point $x \in C$ when there exists a neighborhood U of x such that the intersection $C \cap U$ is a closed set.

Metric Regularity. *A mapping $F : X \rightrightarrows Y$ is said to be metrically regular at $\bar{x}$ for $\bar{y}$ when $\bar{y} \in F(\bar{x})$, the graph of F is locally closed at $(\bar{x}, \bar{y})$, and there is a constant $\kappa > 0$ together with neighborhoods U of $\bar{x}$ and V of $\bar{y}$ such that*

$$d(x, F^{-1}(y)) \leq \kappa d(y, F(x)) \quad \textit{for all } (x, y) \in U \times V. \tag{1}$$

The infimum of κ over all such combinations of κ, U, and V is called the regularity modulus for F at $\bar{x}$ for $\bar{y}$ and denoted by $\operatorname{reg}(F; \bar{x}\,|\,\bar{y})$. *The absence of metric regularity is signaled by* $\operatorname{reg}(F; \bar{x}\,|\,\bar{y}) = \infty$.

For a set-valued mapping $F : X \rightrightarrows Y$ and a point $y \in Y$, metric regularity gives an estimate for how far a point x is from being a solution to the inclusion $F(x) \ni y$ in terms of the "residual" $d(y, F(x))$. To be specific, let $\bar{x}$ be a solution of the inclusion $\bar{y} \in F(x)$, let F be metrically regular at $\bar{x}$ for $\bar{y}$, and let x_{a} and y_{a} be approximations to $\bar{x}$ and $\bar{y}$, respectively. Then from (1), the distance from x_{a} to the set of solutions of the inclusion $y_{\mathrm{a}} \in F(x)$ is bounded by the constant κ times the residual, i.e., the quantity measuring how much y_{a} is away from $F(x_{\mathrm{a}})$. In applications, the residual is typically easy to compute or estimate, whereas finding a solution might be considerably more difficult. Metric regularity says that there exists a solution to the inclusion $y_{\mathrm{a}} \in F(x)$ at distance from x_{a} proportional to the residual. In this way we can obtain an estimate for the distance between approximate solutions to an exact one. In this book we will apply this observation several times in various situations.

For a linear and bounded mapping A acting between Banach spaces X and Y, metric regularity is equivalent to the surjectivity of A, that is, the property $AX = Y$.

© The Author(s), under exclusive license to Springer Nature Switzerland AG 2021
A. L. Dontchev, *Lectures on Variational Analysis*, Applied Mathematical Sciences 205, https://doi.org/10.1007/978-3-030-79911-3_4

This follows from a fundamental result in functional analysis, the Banach open mapping principle. We will state next the following version of it:

Banach Open Mapping Principle. *Let X and Y be Banach spaces. For any linear and bounded single-valued mapping $A : X \to Y$, the following properties are equivalent*:

(a) *A is surjective*;

(b) *A is open (at every point)*;

(c) $0 \in \operatorname{int} A(\operatorname{int} I\!B)$;

(d) *there is a $\kappa > 0$ such that for all $y \in Y$ there exists $x \in X$ with $Ax = y$ and* $\|x\| \leq \kappa\|y\|$.

We will not present here a separate proof of the Banach principle; we will deduce it from a more general result about metric regularity of set-valued mappings with closed and convex graphs, the Robinson–Ursescu theorem, stated and proved in Lecture 6.

Recall that a function $f : X \to Y$ is *open* when at $\bar{x} \in \operatorname{dom} f$ when for any neighborhood U of $\bar{x}$ the set $f(U)$ is a neighborhood of $f(\bar{x})$. This definition and the linearity of A immediately show the equivalence of (a) and (b) above. We will now connect condition (d) with metric regularity.

The first observation to make is that (d) above implies the existence of a $\kappa > 0$ such that $d(0, A^{-1}(y)) \leq \kappa\|y\|$ for all y. The converse also holds; indeed, in that case one may only need to take a slightly bigger κ. From the linearity of A, for $x \in X$ and $y \in Y$ in general, we have $d(x, A^{-1}(y)) = d(0, A^{-1}(y) - x)$. Then since $z \in A^{-1}(y) - x$ corresponds to $A(x+z) = y$, we get $d(0, A^{-1}(y) - x) = d(0, A^{-1}(y - Ax)) \leq \kappa\|y - Ax\|$. Thus, (d) above is equivalent to:

$$\text{there exists } \kappa > 0 \text{ such that } d(x, A^{-1}(y)) \leq \kappa d(y, Ax) \quad \text{for all } x \in X,\ y \in Y.$$

Obviously, this is the same as the metric regularity property as specialized to A. Note that the local property of metric regularity becomes global through the arbitrary scaling made available because $A(\lambda x) = \lambda Ax$. In fact, due to linearity, metric regularity of A with respect to any pair $(\bar{x}, \bar{y})$ in its graph is identical to metric regularity with respect to $(0, 0)$, and the same modulus of metric regularity prevails everywhere. We denote this modulus by $\operatorname{reg} A$.

More could be said here regarding the equivalence of global metric regularity with Lipschitz continuity of the inverse. Indeed, let us prove that (d) is the same as A^{-1} being Lipschitz continuous everywhere with a constant κ. Let $y, y' \in Y$ and $Ax = y$. Then according to (d) there exists $x' \in X$ such that $A(x' - x) = y' - y$, which is the same as $Ax' = y'$, and $\|x' - x\| \leq \kappa\|y' - y\|$. Hence,

$$d(x, A^{-1}y') \leq \|x - x'\| \leq \kappa\|y' - y\|$$

and since the right side of this last inequality does not depend on x, we can take supremum with respect to $x \in A^{-1}(y)$ obtaining that A^{-1} is Lipschitz continuous on Y.

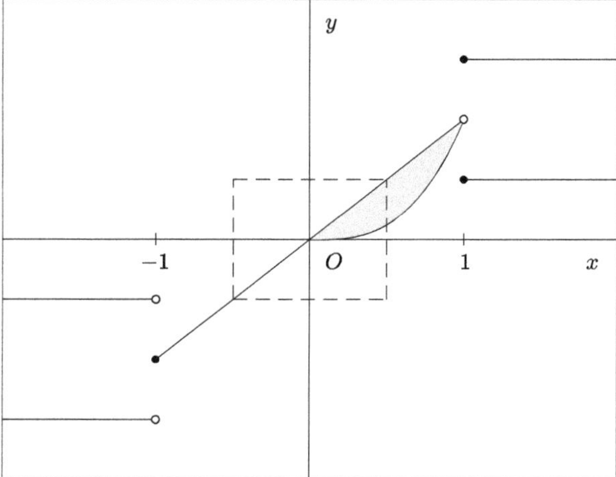

Fig. 4.1: Graph of a mapping that has the Aubin property at 0 for 0 is inner but not outer semicontinuous at −1 and is inner but not outer semicontinuous at 1

Metric regularity of a mapping is equivalent to a property of its inverse that resembles but is weaker than the Lipschitz continuity. We introduce it next for set-valued mappings acting between metric spaces Y and X.

Aubin Property. *A mapping $S : Y \rightrightarrows X$ is said to have the Aubin property at $\bar{y} \in Y$ for $\bar{x} \in X$ if $\bar{x} \in S(\bar{y})$, the graph of S is locally closed at $(\bar{y}, \bar{x})$, and there is a constant $\kappa \geq 0$ together with neighborhoods U of $\bar{x}$ and V of $\bar{y}$ such that*

$$e(S(y') \cap U, S(y)) \leq \kappa\rho(y', y) \quad \textit{for all } \ y', y \in V. \tag{2}$$

Moreover, the infimum of κ for which (2) holds denoted by $\operatorname{lip}(S; \bar{y} \,|\, \bar{x})$. *Writing* $\operatorname{lip}(S; \bar{y} \,|\, \bar{x}) = \infty$ *means that S does not have the Aubin property at $\bar{y}$ for $\bar{x}$.*

In contrast to Lipschitz continuity where $U = X$, the Aubin property is tied to a particular point in the graph of the mapping (see Fig. 4.1). Also, observe that when a set-valued mapping S has the Aubin property at $\bar{y}$ for $\bar{x}$, then, for every point $(y, x) \in \operatorname{gph} S$ that is sufficiently close to $(\bar{y}, \bar{x})$, it has the Aubin property at y for x as well. It is also important to note that the Aubin property of S at $\bar{y}$ for $\bar{x}$ implicitly requires $\bar{y}$ be an element of int dom S; this is exhibited in the following proposition.

Proposition 4.1. *If $S : Y \rightrightarrows X$ has the Aubin property at $\bar{y}$ for $\bar{x}$, then for every neighborhood $\bar{U}$ of $\bar{x}$ there exists a neighborhood $\bar{V}$ of $\bar{y}$ such that $S(y) \cap \bar{U} \neq \emptyset$ for all $y \in \bar{V}$.*

Proof. Let (2) hold with neighborhoods U of $\bar{x}$ and V of $\bar{y}$ and a constant κ. Take any $y \in V$ with $y \neq \bar{y}$; then (2) with $y' = \bar{y}$ becomes

$$e(S(\bar{y}) \cap U, S(y)) \leq \kappa\rho(y, \bar{y}).$$

Take any $\kappa' > \kappa$. Then

$$d(\bar{y}, S(y)) \le \kappa\rho(y, \bar{y}) < \kappa'\rho(y, \bar{y}).$$

This means that the intersection of the ball centered at $\bar{x}$ with radius $\kappa'\rho(y, \bar{y})$ with $S(y)$ is nonempty. Take any neighborhood $\bar{U}$ of $\bar{x}$. Then there exists a neighborhood $\bar{V}$ of $\bar{y}$ such that $I\!B_{\kappa'\rho(y,\bar{y})}(\bar{x}) \subset \bar{U}$ for every $y \in \bar{V}$. But then $S(y) \cap \bar{U} \neq \emptyset$ for all $y \in \bar{V}$. □

The Aubin property could alternatively be defined with one variable "free," as shown in the next proposition.

Proposition 4.2. *A mapping* $S : Y \rightrightarrows X$ *has the Aubin property at* $\bar{y}$ *for* $\bar{x}$ *with constant* $\kappa \ge 0$ *if and only if* $\bar{x} \in S(\bar{y})$ *and there exist neighborhoods* U *of* $\bar{x}$ *and* V *of* $\bar{y}$ *such that*

$$e(S(y') \cap U, S(y)) \le \kappa\rho(y', y) \quad \textit{for all } \; y' \in Y \; \textit{ and } \; y \in V. \tag{3}$$

Proof. Clearly, (3) implies (2). Assume (2) with corresponding U and V and choose positive a and b such that $I\!B_a(\bar{x}) \subset U$ and $I\!B_b(\bar{y}) \subset V$. Let $0 < a' < a$ and $0 < b' < b$ be such that

$$2\kappa b' + a' \le \kappa b. \tag{4}$$

For any $y \in I\!B_{b'}(\bar{y})$ we have from (2) that

$$d(\bar{x}, S(y)) \le \kappa\rho(y, \bar{y}) \le \kappa b',$$

hence

$$e(I\!B_{a'}(\bar{x}), S(y)) \le \kappa b' + a'. \tag{5}$$

Take any $y' \in Y$. If $y' \in I\!B_b(\bar{y})$ the inequality in (3) comes from (2) and there is nothing more to prove. Assume $\rho(y', \bar{y}) > b$. Then $\rho(y, y') > b - b'$ and from (4), $\kappa b' + a' \le \kappa(b - b') \le \kappa\rho(y, y')$. Using this in (5) we obtain

$$e(I\!B_{a'}(\bar{x}), S(y)) \le \kappa\rho(y', y),$$

and since $S(y') \cap I\!B_{a'}(\bar{x})$ is obviously a subset of $I\!B_{a'}(\bar{x})$, we come to (3). □

The following theorem shows the equivalence of metric regularity of a mapping with the Aubin property of the inverse of it.

Theorem 4.3. *A set-valued mapping* $F : X \rightrightarrows Y$ *is metrically regular at* $\bar{x}$ *for* $\bar{y}$ *with a constant* $\kappa > 0$ *if and only if its inverse* F^{-1} *has the Aubin property at* $\bar{y}$ *for* $\bar{x}$ *with constant* $\kappa > 0$*. In this case*

$$\operatorname{reg}(F; \bar{x}\,|\,\bar{y}) = \operatorname{lip}(F^{-1}; \bar{y}\,|\,\bar{x}).$$

Proof. First note that the local closedness of the graph of F at $(\bar{x}, \bar{y})$ is equivalent to the local closedness of the graph of F^{-1} at $(\bar{y}, \bar{x})$. Let $\kappa >$ $\operatorname{reg}(F; \bar{x}\,|\,\bar{y})$; then there are positive constants a and b such that (1) holds with $U = \mathbb{B}_a(\bar{x})$, $V = \mathbb{B}_b(\bar{y})$, and with this κ. Without loss of generality, assume $b < \kappa/a$. Choose $y, y' \in \mathbb{B}_b(\bar{y})$. If $F^{-1}(y) \cap \mathbb{B}_a(\bar{x}) = \emptyset$, then $d(\bar{x}, F^{-1}(y)) \geq a$. But then (1) with $x = \bar{x}$ yields

$$a \leq d(\bar{x}, F^{-1}(y)) \leq \kappa d(y, F(\bar{x})) \leq \kappa\rho(y, \bar{y}) \leq \kappa b < a,$$

a contradiction. Hence, there exists $x \in F^{-1}(y) \cap \mathbb{B}_a(\bar{x})$, and for any such x we have from (1) that

$$d(x, F^{-1}(y')) \leq \kappa d(y', F(x)) \leq \kappa\rho(y', y).$$

Taking the supremum with respect to $x \in F^{-1}(y) \cap \mathbb{B}_a(\bar{x})$, we obtain (3) with $S = F^{-1}$, $U = \mathbb{B}_a(\bar{x})$, and $V = \mathbb{B}_b(\bar{y})$, and therefore, the infimum of κ such that (3) holds is greater than $\operatorname{reg}(F; \bar{x} \,|\, \bar{y})$.

Conversely, suppose there are neighborhoods U of $\bar{x}$ and V of $\bar{y}$ along with a constant κ such that (3) holds with $S = F^{-1}$. Take U and V smaller if necessary so that, according to Proposition 4.2, we have

$$e(F^{-1}(y') \cap U, F^{-1}(y)) \leq \kappa\rho(y', y) \quad \text{for all } y' \in Y \text{ and } y \in V. \tag{6}$$

Let $x \in U$ and $y \in V$. If $F(x) \neq \emptyset$, then for any $y' \in F(x)$, we have $x \in F^{-1}(y') \cap U$. From (6), we obtain

$$d(x, F^{-1}(y)) \leq e(F^{-1}(y') \cap U, F^{-1}(y)) \leq \kappa\rho(y', y).$$

This holds for any $y' \in F(x)$; hence, by taking the infimum with respect to $y' \in F(x)$ in the last expression, we get

$$d(x, F^{-1}(y)) \leq \kappa d(y, F(x)).$$

(If $F(x) = \emptyset$, then because of the convention $d(y, \emptyset) = \infty$, this inequality holds automatically.) Hence, F is metrically regular at $\bar{x}$ for $\bar{y}$ with a constant κ. Then we have that the infimum of κ such that (3) holds is not greater than $\operatorname{reg}(F; \bar{x} \,|\, \bar{y})$. Hence, that infimum equals the regularity modulus $\operatorname{reg}(F; \bar{x} \,|\, \bar{y})$. □

The following theorem shows that for convex-valued mappings, the Aubin property implies that a truncation of the mapping is Lipschitz continuous around the reference point. It is stated and proved for X and Y being Banach spaces but can be easily extended for mappings acting from a metric space to a linear metric space with a shift-invariant metric.

Theorem 4.4. *Let X and Y be Banach spaces. A set-valued mapping $S : Y \rightrightarrows X$, whose graph is locally closed at $(\bar{y}, \bar{x})$ and whose values are convex sets, has the Aubin property at $\bar{y}$ for $\bar{x}$ if and only if it has a Lipschitz continuous graphical localization (not necessarily single-valued) around $\bar{y}$ for $\bar{x}$, or in other words, there are a closed neighborhood U of $\bar{x}$ and a neighborhood V of $\bar{y}$ such that the truncated mapping $y \mapsto S(y) \cap U$ is Lipschitz continuous on V.*

It is important to note that convexifying the values of a set-valued mapping may change significantly the Lipschitz properties of the mapping. As a simple example consider the solution mapping s of the inverse of the function $x \mapsto x^2$ considered in the preparatory section, see Fig. 0.1, and then the mapping

$$\mathbb{R} \ni y \mapsto S(y) = \{s(y)\} \cup \{1\} \cup \{-1\}.$$

The mapping S is not Lipschitz continuous on any interval $[0, a]$, $a > 0$, while $y \mapsto \text{co } S(y)$ is Lipschitz continuous on $\mathbb{R}$.

There is a third property that is closely related to both metric regularity and the Aubin property. At this point we assume that X and Y are Banach spaces, but this is not at all a necessary restriction for all the analysis given below. First, we give a definition.

Openness of a Set-Valued Mapping. *A mapping $F : X \rightrightarrows Y$ is said to be open at $\bar{x}$ for $\bar{y}$ if $\bar{y} \in F(\bar{x})$ and for every neighborhood U of $\bar{x}$, $F(U) := \cup_{x \in U} F(x)$ is a neighborhood of $\bar{y}$.*

It turns out that metric regularity implies openness; it is actually equivalent to the following stronger version of the openness property:

Linear Openness. *A mapping $F : X \rightrightarrows Y$ is said to be linearly open at $\bar{x}$ for $\bar{y}$ when $\bar{y} \in F(\bar{x})$, the graph of F is locally closed at $\bar{x}$ for $\bar{y}$, and there is a constant $\kappa > 0$ together with neighborhoods U of $\bar{x}$ and V of $\bar{y}$ such that*

$$F(x + \kappa r \,\text{int}\, \mathbb{B}) \supset \left[F(x) + r \,\text{int}\, \mathbb{B}\right] \cap V \quad \textit{for all } x \in U \textit{ and all } r > 0. \tag{7}$$

Linear openness is a particular case of openness that is obtained from (7) for $x = \bar{x}$. Linear openness postulates openness *around* the reference point with balls having proportional radii. The following theorem establishes the equivalence of linear openness and metric regularity.

Theorem 4.5. *A set-valued mapping $F : X \rightrightarrows Y$ is linearly open at $\bar{x}$ for $\bar{y}$ if and only if F is metrically regular at $\bar{x}$ for $\bar{y}$. In this case the infimum of κ for which* (7) *holds is equal to* $\text{reg}\,(F; \bar{x}\,|\,\bar{y})$.

Proof. Choose $y \in V$ and $x' \in U$. Let $y' \in F(x')$ (if there is no such y' there is nothing to prove). Since $y = y' + \|y - y'\|w$ for some $w \in \mathbb{B}$, denoting $r = \|y - y'\|$, for every $\varepsilon > 0$, we have $y \in (F(x') + r(1 + \varepsilon)\,\text{int}\, \mathbb{B}) \cap V$. From (7), there exists $x \in F^{-1}(y)$ with

$$\|x - x'\| \le \kappa(1 + \varepsilon)r = \kappa(1 + \varepsilon)\|y' - y\|.$$

Then

$$d(x', F^{-1}(y)) \le \kappa(1 + \varepsilon)\|y' - y\|.$$

Taking infimum with respect to $y' \in F(x')$ on the right and passing to zero with ε (since the left side does not depend on ε), we obtain that F is metrically regular at $\bar{x}$ for $\bar{y}$ with constant κ.

For the converse, we use Proposition 4.2. Let $x \in U$, $r > 0$, and let $y' \in (F(x) + r \operatorname{int} I\!B) \cap V$. Then there exists $y \in F(x)$ such that $\|y - y'\| < r$. If $y = y'$, then $y' \in F(x) \subset F(x + \kappa r \operatorname{int} I\!B)$, which yields (5) with constant κ. Let $y \neq y'$ and let $\varepsilon > 0$ be so small that $(\kappa + \varepsilon)\|y - y'\| < \kappa r$. Using (3) we obtain

$$d(x, F^{-1}(y')) \leq \kappa\|y - y'\| \leq (\kappa + \varepsilon)\|y - y'\|.$$

Then there exists $x' \in F^{-1}(y')$ such that $|x - x'| \leq (\kappa + \varepsilon)\|y - y'\|$. But then

$$y' \in F(x') \subset F(x + (\kappa + \varepsilon)\|y - y'\| I\!B) \subset F(x + \kappa r \operatorname{int} I\!B),$$

which again yields (7) with constant κ. □

Metric regularity, as well as the Aubin property and linear openness, can be defined without requiring local closedness of the graph, and the equivalence among these properties would still be valid. This would trigger relatively easy modifications of various results throughout the book; such modifications are commented in the next lecture for the case of the Lyusternik–Graves theorem. In this book we prefer to keep the local closedness of the graph in the definitions, in order to simplify the presentation.

Metric regularity of a mapping is not inherited in general by metric regularity of a restriction of that mapping on a subspace. This observation is demonstrated by the following example of a mapping acting from $I\!R^2$ to $I\!R$.

Example. The function $f(x_1, x_2) = x_2 - x_1^3$ is metrically regular at $(0, 0)$ for 0. Indeed, for any $x = (x_1, x_2) \in I\!R^2$ and $y \in I\!R$ near the origins $(0, 0)$ and 0 we have

$$d(x, f^{-1}(y)) = \left\| \begin{pmatrix} x_1 \\ x_2 \end{pmatrix} - \begin{pmatrix} x_1 \\ y + x_1^3 \end{pmatrix} \right\| = \|y - (x_2 - x_1^3)\| = \|y - f(x)\|.$$

For the restriction $\tilde{f}$ of f on $\{(x_1, x_2) \mid x_2 = 0\}$ we have $\tilde{f}^{-1}(y) = (-y)^{1/3}$, which is not Lipschitz continuous around 0.

Exercise 4.6. *For metric space X and P consider the optimization problem*

$$\textit{minimize } g(p, x) \quad \textit{subject to} \quad x \in D(p), \tag{8}$$

where $p \in P$ is a parameter, $g : P \times X \to I\!R$, and $D : P \rightrightarrows X$. Assume that the mapping D has the Aubin property at $\bar{p}$ for $\bar{x}$ with neighborhoods U of $\bar{x}$ and Q of $\bar{p}$ and for every $p \in Q$ problem (8) *has a solution in U. Moreover, suppose that g is Lipschitz continuous on the set $Q \times U$. Show that the optimal value S_{val} of* (8) *is Lipschitz continuous in a neighborhood of $\bar{p}$.*

Guide. Let $p', p \in Q$ and let $x(p') \in D(p') \cap U$ be a solution of problem (8). From the Aubin property of D there exists $x \in D(p)$ such that $\rho(x, x(p')) \leq \kappa\rho(p, p')$ for some $\kappa > 0$. Make Q smaller if necessary so that $x \in U$. For the rest, follow the proof of Theorem 2.2. □

Proof of Theorem 4.4.

The "if" part holds even without the convexity assumption. Indeed, if $y \mapsto S(y) \cap U$ is Lipschitz continuous on V, moreover we have

$$S(y') \cap U \subset S(y) \cap U + \kappa|y' - y|\mathbb{B} \subset S(y) + \kappa|y' - y|\mathbb{B} \quad \text{for all } y', y \in V,$$

that is, S has the desired Aubin property. For the "only if" part, suppose now that S has the Aubin property at $\bar{y}$ for $\bar{x}$ with constant $\kappa > 0$, and let $a > 0$ and $b > 0$ be such that

$$S(y') \cap \mathbb{B}_a(\bar{x}) \subset S(y) + \kappa\|y' - y\|\mathbb{B} \quad \text{for all } y', y \in \mathbb{B}_b(\bar{y}), \tag{9}$$

and the set $S(y) \cap \mathbb{B}_a(\bar{x})$ is closed for all $y \in \mathbb{B}_b(\bar{y})$. Adjust a and b so that, by Proposition 4.1,

$$S(y) \cap \mathbb{B}_{a/2}(\bar{x}) \neq \emptyset \text{ for all } y \in \mathbb{B}_b(\bar{x}) \text{ and } b < \frac{a}{4\kappa}. \tag{10}$$

Pick $y, y' \in \mathbb{B}_b(\bar{y})$ and let $x' \in S(y') \cap \mathbb{B}_a(\bar{x})$. Then from (9) there exists $x \in S(y)$ such that

$$\|x - x'\| \leq \kappa\|y - y'\|. \tag{11}$$

If $x \in \mathbb{B}_a(\bar{x})$, there is nothing more to prove, so assume that $r := \|x - \bar{x}\| > a$. By (10), we can choose a point $\tilde{x} \in S(y) \cap \mathbb{B}_{a/2}(\bar{x})$. Since $S(y)$ is convex, there exists a point $z \in S(y)$ on the segment $[x, \tilde{x}]$ such that $\|z - \bar{x}\| = a$ and then $z \in S(y) \cap \mathbb{B}_a(\bar{x})$. We will now show that

$$\|z - x'\| \leq 5\kappa\|y - y'\|, \tag{12}$$

which yields that the mapping $y \mapsto S(y) \cap \mathbb{B}_a(\bar{x})$ is Lipschitz continuous on $\mathbb{B}_b(\bar{y})$ with constant 5κ.

By construction, there exists $t \in (0, 1)$ such that $z = (1 - t)x + t\tilde{x}$. Then

$$a = \|z - \bar{x}\| = \|(1 - t)(x - \bar{x}) + t(\tilde{x} - \bar{x})\| \leq (1 - t)r + t\|\tilde{x} - \bar{x}\|$$

and in consequence $t(r - \|\tilde{x} - \bar{x}\|) \leq r - a$. Since $\tilde{x} \in \mathbb{B}_{a/2}(\bar{x})$, we get

$$t \leq \frac{r - a}{r - a/2}.$$

Using the triangle inequality $\|\tilde{x} - x\| \leq \|x - \bar{x}\| + \|\tilde{x} - \bar{x}\| \leq r + a/2$, we obtain

$$\|z - x\| = t\|\tilde{x} - x\| \leq \frac{r - a}{r - a/2}(r + a/2). \tag{13}$$

Also, in view of (10) and (11), we have that

$$r = \|x - \bar{x}\| \leq \|x' - x\| + \|x' - \bar{x}\| \leq \kappa\|y - y'\|| + a \leq \kappa 2b + a \leq \frac{3a}{2}. \tag{14}$$

From (13), (14), and since $r > a$, we obtain

$$\|z - x\| \leq (r - a)\frac{r + a/2}{r - a/2} \leq (r - a)\frac{3a/2 + a/2}{a - a/2} = 4(r - a). \tag{15}$$

Note that $d := r - a$ is exactly the distance from x to the ball $I\!B_a(\bar{x})$, hence $d \leq \|x - x'\|$ because $x' \in I\!B_a(\bar{x})$. Combining this with (15) and taking into account (11), we arrive at

$$\|z - x'\| \leq \|z - x\| + \|x - x'\| \leq 4d + \|x' - x\| \leq 5\|x - x'\| \leq 5\kappa\|y - y'\|.$$

But this is (6), and we are done. □

Lecture 5
Lyusternik–Graves Theorem

In this lecture we present a fundamental result in variational analysis, which has its roots in two separate results obtained independently by L. A. Lyusternik (1934) and L. M. Graves (1950) and which nowadays is known as the Lyusternik–Graves theorem. Adapted to our terminology, this theorem is stated in the following way:

Theorem 5.1 (Lyusternik–Graves). *Consider a function $f : X \to Y$ acting between Banach spaces X and Y and suppose that f is strictly Fréchet differentiable at $\bar{x} \in \mathrm{dom}\, f$ with the strict derivative $Df(\bar{x})$ being surjective. Then f is metrically regular at $\bar{x}$ for $f(\bar{x})$.*

According to the Banach open mapping principle stated in the preceding Lecture 4, the surjectivity of a linear and bounded mapping is equivalent to its metric regularity at any point. Then the Lyusternik–Graves theorem can be restated as follows: metric regularity of the linearized mapping $x \mapsto F(x) := f(\bar{x}) + Df(\bar{x})(x - \bar{x})$ at $\bar{x}$ for $f(\bar{x})$ implies metric regularity of the function f at $\bar{x}$ for $f(\bar{x})$. Observe that f can be viewed as a "perturbation" of F of the form $f = F + g$, where the perturbation $g(x) = -f(x) + f(\bar{x}) + Df(\bar{x})(x - \bar{x})$ has Lipschitz modulus zero. It turns out that the Lyusternik–Graves theorem is a particular case of the following general paradigm: *For any set-valued mapping F that is metrically regular at $\bar{x}$ for $\bar{y}$, and for any function g, if*

$$\mathrm{lip}\,(g; \bar{x}) \cdot \mathrm{reg}\,(F; \bar{x}\,|\,\bar{y}) < 1,$$

then the mapping $g + F$ is metrically regular at $\bar{x}$ for $g(\bar{x}) + \bar{y}$.

In the pages of the book that follow we will show that this paradigm is also valid for regularity properties of mappings other than metric regularity. In the case of metric regularity, it is specified in the following theorem:

Theorem 5.2. *Let X be a complete metric space and let Y be a linear metric space with shift-invariant metric. Consider a set-valued mapping $F : X \rightrightarrows Y$, a point $(\bar{x}, \bar{y}) \in \mathrm{gph}\, F$, and a function $g : X \to Y$ with $\bar{x} \in \mathrm{int\,dom}\, g$. Let κ and μ be nonnegative constants such that*

© The Author(s), under exclusive license to Springer Nature Switzerland AG 2021
A. L. Dontchev, *Lectures on Variational Analysis*, Applied Mathematical Sciences 205, https://doi.org/10.1007/978-3-030-79911-3_5

$$\kappa\mu < 1, \quad \operatorname{reg}(F;\bar{x}\,|\,\bar{y}) \le \kappa \quad \textit{and} \quad \operatorname{lip}(g;\bar{x}) \le \mu.$$

Then

$$\operatorname{reg}(g+F;\bar{x}\,|\,g(\bar{x})+\bar{y}) \le \frac{\kappa}{1-\kappa\mu}.$$

As noted after its statement, Theorem 5.1 is a special case of Theorem 5.2 for $F(x) = f(\bar{x}) + Df(\bar{x})(x-\bar{x})$ and $g(x) = -f(x) + f(\bar{x}) + Df(\bar{x})(x-\bar{x})$. Indeed, it is enough to observe that

$$\operatorname{lip}(f(\cdot) - f(\bar{x}) + Df(\bar{x})(\cdot - \bar{x}); \bar{x}) = 0. \tag{1}$$

Actually, Theorem 5.1 is covered by the following more general corollary of Theorem 5.2, for the proof of which it is enough to use (1):

Corollary 5.3. *Let X and Y be Banach spaces. For a function $f : X \to Y$ that is strictly differentiable at $\bar{x}$ and a set-valued mapping $F : X \rightrightarrows Y$ with $\bar{y} \in f(\bar{x})+F(\bar{x})$,*

$$\operatorname{reg}(f+F;\bar{x}\,|\,\bar{y}) = \operatorname{reg}(f(\bar{x}) + Df(\bar{x})(\cdot - \bar{x}) + F(\cdot);\bar{x}\,|\,\bar{y}).$$

For a function f acting between Banach spaces X and Y that is strictly differentiable at $\bar{x}$, Corollary 5.3 complements Theorem 5.1 in that metric regularity at $\bar{x}$ for $f(\bar{x})$ is actually *equivalent* to surjectivity of $Df(\bar{x})$. In the case $X = Y = \mathbb{R}^n$ this is the same as nonsingularity of the matrix $\nabla f(\bar{x})$.

Theorem 5.2 will be deduced from the more general Theorem 5.5 given in further lines and supplied with a proof that uses the following fixed point theorem for set-valued mappings:

Theorem 5.4. *Let X be a complete metric space with metric ρ, and consider a set-valued mapping $\Phi : X \rightrightarrows X$ and a point $\bar{x} \in X$. Suppose that there exist scalars $a > 0$ and $\gamma \in (0,1)$ such that the set* gph $\Phi \cap (\mathbb{B}_a(\bar{x}) \times \mathbb{B}_a(\bar{x}))$ *is closed and*

(a) *$d(\bar{x}, \Phi(\bar{x})) < a(1-\gamma)$;*

(b) *$e(\Phi(x) \bigcap \mathbb{B}_a(\bar{x}), \Phi(x')) \le \gamma\, \rho(x, x')$ for all $x, x' \in \mathbb{B}_a(\bar{x})$.*

Then Φ has a fixed point in $\mathbb{B}_a(\bar{x})$; that is, there exists $x \in \mathbb{B}_a(\bar{x})$ such that $x \in \Phi(x)$.

For functions, Theorem 5.4 becomes the classical Banach contraction mapping theorem stated next for completeness.

Banach Contraction Mapping Theorem. *Let X be a complete metric space with metric ρ. Consider a point $\bar{x} \in X$ and a function $\Phi : X \to X$ for which there exist scalars $a > 0$ and $\gamma \in [0,1)$ such that:*

(a) *$\rho(\Phi(\bar{x}), \bar{x}) \le a(1-\gamma)$;*

(b) *$\rho(\Phi(x'), \Phi(x)) \le \gamma\rho(x', x)$ for every $x', x \in \mathbb{B}_a(\bar{x})$.*

Then there is a unique $x \in \mathbb{B}_a(\bar{x})$ satisfying $x = \Phi(x)$.

The following theorem is an implicit mapping version of Theorem 5.2.

Theorem 5.5. *Let X be a complete metric space, let Y be a linear metric space with a shift-invariant metric, and let P be a metric space. For a function $f : P \times X \to Y$ and a set-valued mapping $F : X \rightrightarrows Y$, consider the following generalized equation in which the function f depends on a parameter p:*

$$f(p, x) + F(x) \ni 0$$

with solution mapping

$$S(p) = \left\{ x \,\middle|\, f(p, x) + F(x) \ni 0 \right\} \quad \textit{having } \bar{x} \in S(\bar{p}).$$

Let κ and μ be positive constants such that $\kappa\mu < 1$. Let $h : X \to Y$ be a function that satisfies the conditions

$$h(\bar{x}) = f(\bar{p}, \bar{x}) \qquad \textit{and} \qquad \widehat{\operatorname{lip}}_x(f - h; (\bar{p}, \bar{x})) \leq \mu,$$

and suppose that $h + F$ is metrically regular at $\bar{x}$ for 0 with

$$\operatorname{reg}(h + F; \bar{x}\,|\,0) \leq \kappa.$$

Then the mapping S has the Aubin property at $\bar{p}$ for $\bar{x}$, and moreover,

$$\operatorname{lip}(S; \bar{p}\,|\,\bar{x}) \leq \frac{\kappa}{1 - \kappa\mu} \widehat{\operatorname{lip}}_p(f; (\bar{p}, \bar{x})). \tag{2}$$

Theorem 5.2 follows from Theorem 5.5 when $f(p, x) = -p + g(x)$ for $P = Y$ and $g : X \to Y$. Then $S(p) = (g + F)^{-1}(p)$ and the Aubin property for S is equivalent to metric regularity of $g + F$.

If metric regularity were defined without local closedness of the graph, in order to apply Theorem 5.4 in the proof of Theorem 5.5, we would need to require the graph of $h + F$ to be locally closed. This of course would hold if h is continuous and F has a closed graph. In Theorem 5.2 it would be sufficient to assume that F has a closed graph. Similar modifications would be also required later in the book. It is an open question whether local closedness of the graph of $h + F$ is necessary for Theorem 5.5 to hold.

Exercise 5.6. *Let X and Y be Banach spaces. Consider a function $f : X \to Y$ that is continuously Fréchet differentiable in a neighborhood of a point $\bar{x}$ with the derivative mapping $Df(\bar{x})$ surjective. Prove that for every $\varepsilon > 0$ there exists $\delta > 0$ such that*

$$d(x, f^{-1}(f(\bar{x}))) \leq \varepsilon\|x - \bar{x}\| \quad \textit{whenever } x \in (\bar{x} + \ker Df(\bar{x})) \textit{ and } \|x - \bar{x}\| \leq \delta.$$

Hint. Use the metric regularity of f and the Taylor expansion of f about $\bar{x}$. □

Proofs

Proof of Theorem 5.4. By assumption (a) there exists $x^1 \in \Phi(\bar{x})$ such that $\rho(x^1, \bar{x}) < a(1-\gamma)$. Proceeding by induction, let $x^0 = \bar{x}$ and suppose that there exists $x^{k+1} \in \Phi(x^k) \cap I\!B_a(\bar{x})$ for $k = 0, 1, \ldots, j-1$ with

$$\rho(x^{k+1}, x^k) < a(1-\gamma)\gamma^k.$$

By assumption (b),

$$d(x^j, \Phi(x^j)) \leq e(\Phi(x^{j-1}) \cap I\!B_a(\bar{x}), \Phi(x^j)) \leq \gamma\, \rho(x^j, x^{j-1}) < a(1-\gamma)\gamma^j.$$

This implies that there is an $x^{j+1} \in \Phi(x^j)$ such that

$$\rho(x^{j+1}, x^j) < a(1-\gamma)\gamma^j.$$

By the triangle inequality,

$$\rho(x^{j+1}, \bar{x}) \leq \sum_{i=0}^{j} \rho(x^{i+1}, x^i) < a(1-\gamma)\sum_{i=0}^{j} \gamma^i < a.$$

Hence, $x^{j+1} \in \Phi(x^j) \cap I\!B_a(\bar{x})$ and the induction step is complete.

For any $k > m > 1$ we then have

$$\rho(x^k, x^m) \leq \sum_{i=m}^{k-1} \rho(x^{i+1}, x^i) < a(1-\gamma)\sum_{i=m}^{k-1} \gamma^i < a\gamma^m.$$

Thus, $\{x^k\}$ is a Cauchy sequence and hence it converges to some $x \in I\!B_a(\bar{x})$. Since $(x^{k-1}, x^k) \in \operatorname{gph}\Phi \cap (I\!B_a(\bar{x}) \times I\!B_a(\bar{x}))$ that is a closed set, we conclude that $x \in \Phi(x) \cap I\!B_a(\bar{x})$; that is, x is a fixed point of Φ in $I\!B_a(\bar{x})$. □

Proof of Theorem 5.5. For simplicity we denote all metrics by $\rho(\cdot,\cdot)$. If

$$\widehat{\operatorname{lip}}_p(f; (\bar{p}, \bar{x})) = +\infty,$$

the right side of (2) becomes $+\infty$ and there is nothing to prove. Assume that $\widehat{\operatorname{lip}}_p(f; (\bar{p}, \bar{x})) < \gamma$ for some positive γ, that is, f is Lipschitz continuous with respect to p at $(\bar{p}, \bar{x})$ uniformly in x around $\bar{x}$ with Lipschitz constant γ.

Choose $\lambda > \kappa$ and $\nu > \mu$ such that $\lambda\nu < 1$. Then there exist positive scalars α and τ such that $\operatorname{gph}(h+F) \times (I\!B_\alpha(\bar{x}) \times I\!B_\alpha(0))$ is a closed set,

$$e\Big((h+F)^{-1}(y') \cap I\!B_\alpha(\bar{x}), (h+F)^{-1}(y)\Big) \leq \lambda\, \rho(y', y) \quad \text{for all } y', y \in I\!B_\alpha(0), \quad (3)$$

and

$$\rho(r(p,x'),r(p,x)) \le \nu\,\rho(x',x) \quad \text{for all } x',x \in I\!B_\alpha(\bar{x}) \quad \text{and} \quad p \in I\!B_\tau(\bar{p}), \tag{4}$$

where $r(p,x) = f(p,x) - h(x)$, and

$$\rho(f(p',x),f(p,x)) \le \gamma\,\rho(p',p) \quad \text{for all } p',p \in I\!B_\tau(\bar{p}) \quad \text{and} \quad x \in I\!B_\alpha(\bar{x}). \tag{5}$$

Let

$$\frac{2\lambda\gamma}{1-\lambda\nu} \ge \lambda^+ > \frac{\lambda\gamma}{1-\lambda\nu}. \tag{6}$$

Now, choose positive $a < \alpha$ and then positive $q \le \tau$ such that

$$\nu a + \gamma q \le \alpha \quad \text{and} \quad \frac{4\lambda\gamma q}{1-\lambda\nu} + a \le \alpha. \tag{7}$$

Then, from (4) and (5), for every $x \in I\!B_a(\bar{x})$ and $p \in I\!B_q(\bar{p})$, we have

$$\begin{aligned} \rho(r(p,x),0) &\le \rho(r(p,x),r(p,\bar{x})) + \rho(r(p,\bar{x}),r(\bar{p},\bar{x})) \\ &\le \nu\,\rho(x,\bar{x}) + \gamma\,\rho(p,\bar{p}) \le \nu a + \gamma q \le \alpha. \end{aligned} \tag{8}$$

Fix $p \in I\!B_q(\bar{p})$ and consider the mapping

$$\Phi_p : x \mapsto (h+F)^{-1}(-r(p,x)) \quad \text{for } x \in I\!B_\alpha(\bar{x}).$$

Observe that for any $x \in I\!B_a(\bar{x})$ and $p \in I\!B_q(\bar{p})$, $x \in \Phi_p(x) \iff x \in S(p)$ and also that the set gph $\Phi_p \cap (I\!B_\alpha(\bar{x}) \times I\!B_\alpha(\bar{x}))$ is closed. Let $p',p \in I\!B_q(\bar{p})$ with $p \ne p'$ and let $x' \in S(p') \cap I\!B_a(\bar{x})$. Let $\varepsilon := \lambda^+\rho(p',p)$; then from (6), $\varepsilon \le \lambda^+(2q)$. Thus, remembering that $x' \in \Phi_{p'}(x') \cap I\!B_\alpha(\bar{x})$, from (3), where we use (8), and from (5) and (6), we deduce that

$$\begin{aligned} d(x',\Phi_p(x')) &\le e\Big((h+F)^{-1}(-r(p',x')) \cap I\!B_\alpha(\bar{x}), (h+F)^{-1}(-r(p,x'))\Big) \\ &\le \lambda\,\rho(f(p',x'),f(p,x')) \le \lambda\gamma\,\rho(p',p) < \lambda^+(1-\lambda\nu)\rho(p',p) = \varepsilon(1-\lambda\nu). \end{aligned}$$

Since $x' \in I\!B_a(\bar{x})$ and, by (4) and (5),

$$\varepsilon \le 2\lambda^+ q \le \frac{4\lambda\gamma q}{1-\lambda\nu},$$

we get $I\!B_\varepsilon(x') \subset I\!B_\alpha(\bar{x})$. Then, for any $u,v \in I\!B_\varepsilon(x')$ using again (3) and (4), we see that

$$\begin{aligned} &e(\Phi_p(u) \cap I\!B_\varepsilon(x'), \Phi_p(v)) \\ &\quad \le e\Big((h+F)^{-1}(-r(u,p)) \cap I\!B_\alpha(\bar{x}), (h+F)^{-1}(-r(v,p))\Big) \\ &\quad \le \lambda\,\rho(r(p,u),r(p,v)) \le \lambda\nu\,\rho(u,v)\,. \end{aligned}$$

Hence, Theorem 5.4 applies, with γ there taken to be $\lambda\nu$ here, and it follows that there exists $x \in \Phi_p(x) \cap I\!B_\varepsilon(x')$ and hence $x \in S(p) \cap I\!B_\varepsilon(x')$. Thus,

$$d(x', S(p)) \leq \rho(x', x) \leq \varepsilon = \lambda^{+}\rho(p', p).$$

Since this inequality holds for any $x' \in S(p') \cap \mathbb{B}_a(\bar{x})$ and any λ^{+} fulfilling (5), we arrive at

$$e(S(p') \cap \mathbb{B}_a(\bar{x}), S(p)) \leq \lambda^{+}\rho(p', p).$$

That is, S has the Aubin property at $\bar{p}$ for $\bar{x}$ with modulus not greater than λ^{+}. Since λ^{+} can be arbitrarily close to $\lambda/(1-\lambda\nu)$, and λ, ν, and γ can be arbitrarily close to κ, μ, and $\widehat{\text{lip}}_p(f;(\bar{p},\bar{x}))$, respectively, the proof is complete. □

Lecture 6
Mappings with Convex Graphs

This lecture is devoted to a far reaching generalization of the Banach open mapping principle for mappings whose graphs are closed and convex sets. The simplest example of such a mapping is any linear and bounded mapping $A : X \to Y$; indeed, the graph of such a mapping is a closed linear subspace in $X \times Y$. More generally, any mapping of the form $A + C$, where $A : X \to Y$ is linear and bounded and $C \subset Y$ is a closed convex set and has closed and convex graph. Here and throughout this lecture X and Y are Banach spaces.

The main result in this lecture, proved independently by S. M. Robinson (1976) and C. Ursescu (1975), known as the Robinson–Ursescu theorem, gives necessary and sufficient conditions for metric regularity of a set-valued mapping with closed and convex graph. It will be presented after a series of four propositions whose proofs are given at the end of the lecture.

Observe that for any mapping $F : X \rightrightarrows Y$ with convex graph, the sets dom F and rge F, which are the projections of gph F on the spaces X and Y, are convex sets as well. When gph F is closed, these sets can fail to be closed. The first proposition shows that, when the domain of F is bounded, the interior of the range of F coincides with the interior of its closure.

Proposition 6.1. *For any mapping $F : X \rightrightarrows Y$ with closed convex graph and bounded domain,*

$$\operatorname{int}\operatorname{cl}\operatorname{rge} F = \operatorname{int}\operatorname{rge} F.$$

For the next proposition, we need a definition:

Core and Absorbing Set. *Given a set $C \subset X$, the set containing every x such that for any $w \in X$ there exists $\varepsilon > 0$ such that $tw \in C$ for all t with $0 \le t \le \varepsilon$ is said to be the core of C. A set C is called absorbing if $0 \in$ core C.*

Proposition 6.2. *Let C be a closed convex set in X that is absorbing. Then $0 \in \operatorname{int} C$.*

© The Author(s), under exclusive license to Springer Nature Switzerland AG 2021

A. L. Dontchev, *Lectures on Variational Analysis*, Applied Mathematical Sciences 205, https://doi.org/10.1007/978-3-030-79911-3_6

As we will show in the proof given at the end of the lecture, this proposition is a direct consequence of a fundamental result in analysis/topology, the Baire category theorem, a version of which we state next.

Baire Theorem. *Let X be a complete metric space, and let $\{C_n\}$ be a sequence of closed subsets of X. If* int C_n *is empty for all n, then* int $\bigcup_n C_n$ *is also empty.*

The definition of openness of a mapping $F : X \rightrightarrows Y$ at $\bar{x}$ for $\bar{y}$ in Lecture 4 can be restated equivalently in the following way:

$$\text{for any } a > 0 \text{ there exists } b > 0 \text{ such that } F(\bar{x} + a \operatorname{int} I\!B) \supset \bar{y} + b \operatorname{int} I\!B. \tag{1}$$

Recall that linear openness requires a linear scaling relationship between a and b. The following proposition shows that an intermediate type of property holds automatically when the graph of F is convex.

Proposition 6.3. *Consider a mapping $F : X \rightrightarrows Y$ with convex graph, and let $\bar{y} \in F(\bar{x})$. Then openness of F at $\bar{x}$ for $\bar{y}$ as defined in* (1) *is equivalent to the simpler condition that*

$$\textit{there exists } c > 0 \textit{ with } F(\bar{x} + \operatorname{int} I\!B) \supset \bar{y} + c \operatorname{int} I\!B. \tag{2}$$

The next proposition bridges between condition (2) and metric regularity of set-valued mappings with convex graphs.

Proposition 6.4. *Let $F : X \rightrightarrows Y$ have convex graph containing $(\bar{x}, \bar{y})$, and suppose that condition* (2) *is fulfilled. Then*

$$d(x, F^{-1}(y)) \le \frac{1 + \|x - \bar{x}\|}{c - \|y - \bar{y}\|} d(y, F(x)) \quad \textit{for all } x \in X,\ y \in \bar{y} + c \operatorname{int} I\!B. \tag{3}$$

Now we are ready to state and prove the following version of the Robinson–Ursescu theorem:

Theorem 6.5 (Robinson–Ursescu). *Let $F : X \rightrightarrows Y$ have closed convex graph and let $\bar{y} \in F(\bar{x})$. Then the following are equivalent:*

(a) $\bar{y} \in \operatorname{int} \operatorname{rge} F$.
(b) *F is open at $\bar{x}$ for $\bar{y}$.*
(c) *F is metrically regular at $\bar{x}$ for $\bar{y}$.*

Proof. We first demonstrate that

$$\bar{y} \in \operatorname{int} F(\bar{x} + \operatorname{int} I\!B) \quad \text{when } \bar{x} \in F^{-1}(\bar{y}) \text{ and } \bar{y} \in \operatorname{int} \operatorname{rge} F. \tag{4}$$

By a translation, we can reduce to the case of $(\bar{x}, \bar{y}) = (0, 0)$. To obtain (4) in this setting, where $F(0) \ni 0$ and $0 \in \operatorname{int} \operatorname{rge} F$, it will be enough to show that $0 \in \operatorname{int} F(\delta I\!B)$ for some $\delta \in (0, 1)$. Define the mapping $F_\delta : X \rightrightarrows Y$ by $F_\delta(x) = F(x)$

when $x \in \delta I\!B$ but $F_\delta(x) = \emptyset$ otherwise. Then F_δ has closed convex graph given by $[\delta I\!B \times Y] \cap \operatorname{gph} F$. Also $F(\delta I\!B) = \operatorname{rge} F_\delta$ and $\operatorname{dom} F_\delta \subset \delta I\!B$. We want to show that $0 \in \operatorname{int} \operatorname{rge} F_\delta$.

Taking into account Propositions 6.1 and 6.2, it is sufficient to show that $0 \in \operatorname{core} \operatorname{rge} F_\delta$. For that purpose we use an argument that parallels one already presented in the proof of Proposition 6.1. Consider any $y \in Y$. Because $0 \in \operatorname{int} \operatorname{rge} F$, there exists t_0 such that $ty \in \operatorname{rge} F$ when $t \in [0, t_0]$. Then there exists x_0 such that $t_0 y \in F(x_0)$. Let $x = x_0/t_0$, so that $(t_0 x, t_0 y) \in \operatorname{gph} F$. Since $\operatorname{gph} F$ is convex and contains $(0, 0)$, it also then contains (tx, ty) for all $t \in [0, t_0]$. Taking $\varepsilon > 0$ for which $\varepsilon\|x\| \le \delta$, we get for all $t \in [0, \varepsilon]$ that $(tx, ty) \in \operatorname{gph} F_\delta$, hence $ty \in \operatorname{rge} F_\delta$, as desired.

Utilizing (4), we can put the argument for the equivalences of (a), (b), and (c) together. That (b) implies (a) is obvious. We work next on getting from (a) to (c). When (a) holds, we have from (4) that (2) holds for some c, in which case Proposition 6.4 provides (3). By restricting x and y to small neighborhoods of $\bar{x}$ and $\bar{y}$ in (3), we deduce the metric regularity of F at $\bar{x}$ for $\bar{y}$ with any constant $\kappa > 1/c$. Thus, (c) holds. Finally, out of (c) and Theorem 4.5, we conclude that F is linearly open at $\bar{x}$ for $\bar{y}$, and hence it is open, which is (b). □

The preceding argument shows that linear openness can be added to the equivalences in Theorem 6.5.

Corollary 6.6. *For a mapping $F : X \rightrightarrows Y$ with closed convex graph, openness at $\bar{x}$ for $\bar{y}$ is equivalent to linear openness at $\bar{x}$ for $\bar{y}$.*

Exercise 6.7. *Derive from Theorem 6.5 the Banach open mapping principle stated in Lecture 4.*

Guide. It was already noted after the statement of the Banach open mapping principle that condition (d) in that result was equivalent to global metric regularity of the linear mapping A. It remains only to observe that when Theorem 6.5 is applied to $F = A \in \mathcal{L}(X, Y)$ with $\bar{x} = 0$ and $\bar{y} = 0$, the graph of A being a closed subspace of $X \times Y$ (in particular a convex set), and the positive homogeneity of A is brought in, we not only get statements (b) and (c), but also (a) in the Banach open mapping principle. □

In the remainder of this lecture we apply the Robinson–Ursescu Theorem 6.5 to the mapping describing the following system of nonlinear equalities and inequalities:

$$\begin{cases} f_i(x) \le 0 & \text{for } i \in [1, s], \\ f_i(x) = 0 & \text{for } i \in [s+1, m], \end{cases} \tag{5}$$

where $x \in I\!R^n$ and $f_i : I\!R^n \to I\!R, i = 1, \dots, m$. This kind of system usually describes constraints in a nonlinear programming problem. In terms of the set $D = I\!R^s_+ \times \{0\}^{m-s}$ and the function $f(x) = (f_1(x), \dots, f_m(x))$, system (5) can be written as a generalized equation of the form

$$f(x) + D \ni 0.$$

Let $\bar{x}$ be a solution of this system and suppose that each f_i is continuously differentiable around $\bar{x}$. Define the set-valued mapping $x \mapsto F(x) = f(x) + D$. Then $0 \in F(\bar{x})$. We will apply both Lyusternik–Graves and Robinson–Ursescu theorems to show that the mapping F is metrically regular at $\bar{x}$ for 0 if and only if the Mangasarian–Fromovitz condition is satisfied.

We already mentioned the Mangasarian–Fromovitz condition in Lecture 1; let us state it again:

$$\exists\, w \in I\!R^n \quad \text{with} \quad \begin{cases} \nabla f_i(\bar{x})w < 0 & \text{for } i \in [1, s] \text{ with } f_i(\bar{x}) = 0, \\ \nabla f_i(\bar{x})w = 0 & \text{for } i \in [s+1, m], \end{cases} \tag{6}$$
$$\text{and the vectors } \nabla f_i(\bar{x}) \text{ for } i \in [s+1, m] \text{ are linearly independent.}$$

At this point we need some notation. Assuming that (6) holds, denote by $\nabla f^1(\bar{x})$ the submatrix of $\nabla f(\bar{x})$ with rows $\nabla f_i(\bar{x})$ for which $\nabla f_i(\bar{x})w < 0$ for $i \in [1, s]$ with $f_i(\bar{x}) = 0$ and let the number of rows of that submatrix be d_1. Symmetrically, denote by $\nabla f^2(\bar{x})$ the submatrix of $\nabla f(\bar{x})$ with rows $\nabla f_i(\bar{x})$ for $i \in [s+1, m]$ whose number is $d_2 = m - s$.

Consider the mapping $x \mapsto L(x) = f(\bar{x}) + \nabla f(\bar{x})(x - \bar{x}) + D$ obtained by linearizing f at $\bar{x}$. Since this mapping is the sum of a linear mapping with the convex and closed set $D = I\!R^s_+ \times \{0\}^{m-s}$, we have that gph L is closed and convex. We will first show that the Mangasarian–Fromovitz condition implies

$$0 \in \text{int rge } L. \tag{7}$$

On the contrary, assume that (6) holds with some w and that (7) is violated, that is, 0 is on the boundary of L. We will now utilize the following result:

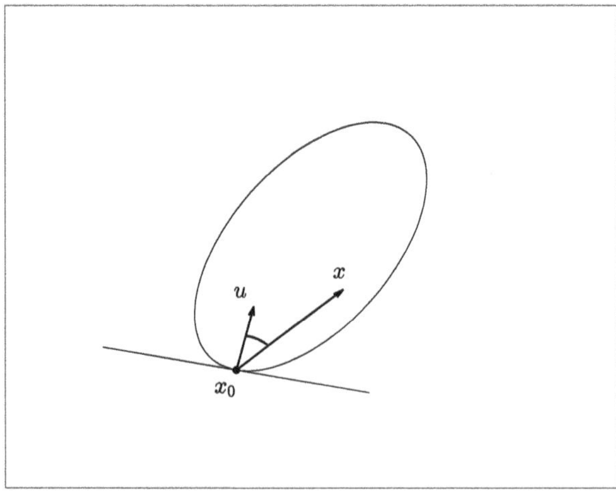

Fig. 6.1: Supporting hyperplane to a convex set

Supporting Hyperplane to a Convex Set. *Let C be a convex set in $\mathbb{R}^n$ and let x_0 be a point on the boundary of C. Then there exists $u \in \mathbb{R}^n$ such that*

$$u^{\mathsf{T}}(x - x_0) \geq 0 \quad \textit{for all} \quad x \in C.$$

This is a special case of the theorem for separation of a convex set from a point outside it by a hyperplane stated in the following lecture (Fig. 6.1).

Coming back to the proof, we conclude that there exists a supporting hyperplane to rge L at 0; that is, there exists a nonzero vector $(u_1, u_2) \in \mathbb{R}^{d_1} \times \mathbb{R}^{d_2}$ such that

$$u_1^{\mathsf{T}} y_1 + u_2^{\mathsf{T}} y_2 \geq 0 \qquad \text{for all } (y_1, y_2) \in \operatorname{rge} L.$$

This means that for all $z \in \mathbb{R}^n$ and $v \in \mathbb{R}^{d_1}_+$

$$u_1^{\mathsf{T}}(\nabla f^1(\bar{x})z + v) + u_2^{\mathsf{T}} \nabla f^2(\bar{x})z \geq 0. \tag{8}$$

For $z = 0$ in (8) we obtain $u_1^{\mathsf{T}} v \geq 0$ for all $v \in \mathbb{R}^{d_1}_+$, hence $u_1 \geq 0$. Taking $z = w$ with w as in (6) and $v = 0$ in (8) we obtain $u_1^{\mathsf{T}} \nabla f^1(\bar{x})w \geq 0$. But $u_1 \geq 0$ and $\nabla f^1(\bar{x})w < 0$ from (6), hence $u_1 = 0$. Then (8) yields $u_2^{\mathsf{T}} \nabla f^2(\bar{x})z \geq 0$ for all $z \in \mathbb{R}^n$, which implies $u_2^{\mathsf{T}} \nabla f^2(\bar{x})z = 0$ for all $z \in \mathbb{R}^n$, and hence, $\nabla f^2(\bar{x})^{\mathsf{T}} u^2 = 0$. But from (6) the rows of $\nabla f^2(\bar{x})$ are linearly independent, hence, ker $\nabla f^2(\bar{x}) = \{0\}$ and then $u_2 = 0$. We obtain that $(u_1, u_2) = 0$, which is a contradiction. Hence, (7) is satisfied.

Applying the Robinson–Ursescu Theorem 6.5, we conclude that the mapping L is metrically regular at 0 for 0. Note that $F(x) = L(x) + h(x)$, where $h(x) = f(x) - f(\bar{x}) - \nabla f(\bar{x})(x - \bar{x})$ has Lipschitz modulus equal to zero. Hence, according to Corollary 5.3 of the (extended) Lyusternik–Graves Theorem 5.2, metric regularity of the mapping L is equivalent to metric regularity of F at $\bar{x}$ for 0. Thus, the Mangasarian–Fromovitz condition implies metric regularity of the mapping F.

We will now show that metric regularity of the mapping F implies the Mangasarian–Fromovitz condition. First, again by Corollary 5.3, it is enough to assume metric regularity of L at 0 for 0, which is the same as openness of L at the same point, according to (Robinson–Ursescu) Theorem 6.5. In the notation used in preceding lines, openness of L means the following: for every $(u^1, u^2) \in \mathbb{R}^{d_1} \times \mathbb{R}^{d_2}$ there exist $w, y \in \mathbb{R}^n$ and $z \in \mathbb{R}^{d_1}$ with $z_i < 0$ such that

$$\begin{cases} \nabla f_i(\bar{x})w - z_i = u_i^1 & \text{for } i \in [1, s] \text{ with } f_i(\bar{x}) = 0, \\ \nabla f_i(\bar{x})(w + y) = u_i^2 & \text{for } i \in [s+1, m]. \end{cases} \tag{9}$$

Taking $u^1 = 0$ in the first line of (9) gives us immediately the first part of the Mangasarian–Fromovitz condition. From the second line of (9) we get that $\nabla f^2(\bar{x})$ must be of full rank, which implies the second part follows. This completes the proof of the equivalence of the Mangasarian–Fromovitz condition and the property of metric regularity of the mapping F.

Proofs

Proof of Proposition 6.1. We will verify that int rge $F \supset$ int cl rge F; this gives equality, inasmuch as the opposite inclusion is obvious. In fact, for this we only need to show that rge $F \supset$ int cl rge F.

Choose $\tilde{y} \in$ int cl rge F; then there exists $\delta > 0$ such that int $\mathbb{B}_{2\delta}(\tilde{y}) \subset$ int cl rge F. We will find a point $\tilde{x}$ such that $(\tilde{x}, \tilde{y}) \in$ gph F, so that $\tilde{y} \in$ rge F. The point $\tilde{x}$ will be obtained as a limit of an infinite sequence (x^k, y^k) that we now construct by induction.

Pick any $(x^0, y^0) \in$ gph F. Suppose we have already determined $(x^j, y^j) \in$ gph F for $j = 0, 1, \ldots, k$. If $y^k = \tilde{y}$, then take $\tilde{x} = x^k$ and $(x^n, y^n) = (\tilde{x}, \tilde{y})$ for all $n = k, k+1, \ldots$; that is, after the index k the sequence is constant. Otherwise, with $\alpha^k := \delta/\|y^k - \tilde{y}\|$, we let $w^k := \tilde{y} + \alpha^k(\tilde{y} - y^k)$. Then $w^k \in \mathbb{B}_\delta(\tilde{y}) \subset$ cl rge F. Hence, there exists $v^k \in$ rge F such that $\|v^k - w^k\| \le \|y^k - \tilde{y}\|/2$ and also u^k with $(u^k, v^k) \in$ gph F. Having gotten this far, we pick

$$(x^{k+1}, y^{k+1}) = \frac{\alpha^k}{1+\alpha^k}(x^k, y^k) + \frac{1}{1+\alpha^k}(u^k, v^k).$$

Clearly, $(x^{k+1}, y^{k+1}) \in$ gph F by its convexity. Also, the sequence y^k satisfies

$$\|y^{k+1} - \tilde{y}\| = \frac{\|v^k - w^k\|}{1+\alpha^k} \le \frac{1}{2}\|y^k - \tilde{y}\|.$$

If $y^{k+1} = \tilde{y}$, we take $\tilde{x} = x^{k+1}$ and $(x^n, y^n) = (\tilde{x}, \tilde{y})$ for all $n = k+1, k+2, \ldots$. If not, we perform the induction step again. As a result, we generate an infinite sequence (x^k, y^k), each element of which is equal to $(\tilde{x}, \tilde{y})$ after some k or has $y^k \ne \tilde{y}$ for all k and also

$$\|y^k - \tilde{y}\| \le \frac{1}{2^k}\|y^0 - \tilde{y}\| \quad \text{for all } k = 1, 2, \ldots. \tag{10}$$

In the latter case, we have $y^k \to \tilde{y}$. Further, for the associated sequence x^k we obtain

$$\|x^{k+1} - x^k\| = \frac{\|x^k - u^k\|}{1+\alpha^k} \le \frac{\|x^k\| + \|u^k\|}{\|y^k - \tilde{y}\| + \delta}\|y^k - \tilde{y}\|.$$

Both x^k and u^k are from dom F and thus are bounded. Therefore, from (10), $\{x^k\}$ is a Cauchy sequence, hence (because X is a complete metric space) convergent to some $\tilde{x}$. Because gph F is closed, we end up with $(\tilde{x}, \tilde{y}) \in$ gph F, as required. □

Proof of Proposition 6.2. Let C be a closed convex set in X that is absorbing. Then $0 \in$ core C, which by the definition of the core implies that for any $x \in X$ there exists n_0 such that $x \in nC$ for all $n \ge n_0$. Thus, we have that $X = \cup_{n=1}^\infty nC$. By Baire theorem, at least one of nC has a nonempty interior. By rescaling, it follows that C has a nonempty interior, that is, there exist $x \in C$ and $\varepsilon > 0$ such that $x + \varepsilon\mathbb{B} \subset C$. Since C is absorbing, taking $\varepsilon > 0$ smaller if necessary, we have that $x + tw \in C$ for all $w \in \mathbb{B}$ and all t with $0 \le t \le \varepsilon$. Moreover, from the assumption that $0 \in$ int core C, taking $x = 0$ and $w = -x$ in the definition of the core, get that $-\delta x \in C$ for some $\delta \in (0, \varepsilon]$. Taking $t = \delta$ we now have that $x + t\mathbb{B} \in C$ and $-tx \in C$. Let $r = t^2/(1+t)$. By the convexity of C, any point in $r\mathbb{B}$ is from C, hence the conclusion. □

Proof of Proposition 6.3. Clearly, (1) implies (2). For the converse, assume (2) and consider any $a > 0$. Take $b = \min\{c, ac\}$. If $a \geq 1$, the left side of (2) is contained in the left side of (1), and hence (1) holds. Suppose therefore that $a < 1$. Let $w \in \bar{y} + b \operatorname{int} \mathbb{B}$. The point $v = (w/a) - (1-a)(\bar{y}/a)$ satisfies $\|v - \bar{y}\| = \|w - \bar{y}\|/a < b/a = c$, hence $v \in \bar{y} + c \operatorname{int} \mathbb{B}$. Then from (2) there exists $u \in \bar{x} + \operatorname{int} \mathbb{B}$ with $(u, v) \in \operatorname{gph} F$. The convexity of $\operatorname{gph} F$ implies $a(u, v) + (1-a)(\bar{x}, \bar{y}) \in \operatorname{gph} F$ and yields

$$av + (1-a)\bar{y} \in F(au + (1-a)\bar{x}) \subset F(\bar{x} + a \operatorname{int} \mathbb{B}).$$

Substituting $v = (w/a) - (1-a)(\bar{y}/a)$ in this inclusion, we see that $w \in F(\bar{x} + a \operatorname{int} \mathbb{B})$, and since w was an arbitrary point in $\bar{y} + b \operatorname{int} \mathbb{B}$, we get (1). □

Proof of Proposition 6.4. We may assume that $(\bar{x}, \bar{y}) = (0, 0)$, since this can be arranged by translating $\operatorname{gph} F$ to $\operatorname{gph} F - (\bar{x}, \bar{y})$. Then condition (2) has the simpler form

$$\textit{there exists } c > 0 \textit{ with } F(\operatorname{int} \mathbb{B}) \supset c \operatorname{int} \mathbb{B}. \tag{11}$$

Let $x \in X$ and $y \in c \operatorname{int} \mathbb{B}$. Observe that (3) is automatically true when $x \notin \operatorname{dom} F$ or $y \in F(x)$, so assume that $x \in \operatorname{dom} F$ but $y \notin F(x)$. Let $\alpha := c - \|y\|$. Then $\alpha > 0$. Choose $\varepsilon \in (0, \alpha)$ and find $y' \in F(x)$ such that $\|y' - y\| \leq d(y, F(x)) + \varepsilon$. The point $\tilde{y} := y + (\alpha - \varepsilon)\|y' - y\|^{-1}(y - y')$ satisfies $\|\tilde{y}\| \leq \|y\| + \alpha - \varepsilon = c - \varepsilon < c$, hence $\tilde{y} \in c \operatorname{int} \mathbb{B}$. By (11) there exists $\tilde{x} \in \operatorname{int} \mathbb{B}$ with $\tilde{y} \in F(\tilde{x})$. Let $\beta := \|y - y'\|(\alpha - \varepsilon + \|y - y'\|)^{-1}$; then $\beta \in (0, 1)$. From the convexity of $\operatorname{gph} F$ we have

$$y = (1-\beta)y' + \beta\tilde{y} \in (1-\beta)F(x) + \beta F(\tilde{x}) \subset F((1-\beta)x + \beta\tilde{x}).$$

Thus $x + \beta(\tilde{x} - x) \in F^{-1}(y)$, so $d(x, F^{-1}(y)) \leq \beta\|x - \tilde{x}\|$. Noting that $\|x - \tilde{x}\| \leq \|x\| + \|\tilde{x}\| < \|x\| + 1$ and $\beta \leq (\alpha - \varepsilon)^{-1}\|y - y'\|$, we obtain

$$d(x, F^{-1}(y)) < \frac{1 + \|x\|}{\alpha - \varepsilon}[d(y, F(x)) + \varepsilon].$$

Letting $\varepsilon \to 0$, we finish the proof. □

Lecture 7
Derivative Criteria for Metric Regularity

In this lecture, we will characterize metric regularity by using generalized derivatives of set-valued mappings. To make things simpler, we limit our considerations to mappings in Euclidean spaces. Some of the results can be extended to infinite dimensions but we will not do that here.

The concept of a tangent cone $T_C(x)$ to a set C in $I\!R^n$ at a point $x \in C$ was introduced in Lecture 1. Recall that a vector $v \in I\!R^n$ is said to be *tangent* to a set $C \subset I\!R^n$ at a point $x \in C$ if

$$\frac{1}{\tau^k}(x^k - x) \to v \quad \text{for some } \; x^k \to x, \; x^k \in C, \; \tau^k \searrow 0.$$

The set of all such vectors v is called the *tangent cone* to C at x and is denoted $T_C(x)$. The *tangent cone mapping* is defined as

$$T_C : x \mapsto \begin{cases} T_C(x) & \text{for } x \in C, \\ \emptyset & \text{otherwise.} \end{cases}$$

The tangent cone $T_C(x)$ is always a closed set. When C is a smooth manifold in $I\!R^n$, $T_C(x)$ is the usual tangent subspace. When the set C is convex, the tangent cone $T_C(x)$ is also convex for any $x \in C$.

In elementary calculus, derivatives are illustrated in terms of tangents to the graph of a function. Let $f : I\!R^n \to I\!R^m$ be a function which is differentiable at x with derivative $\nabla f(x) : I\!R^n \to I\!R^m$. Then

$$(u, v) \in \operatorname{gph} \nabla f(x) \iff (u, v) \in T_{\operatorname{gph} f}(x, f(x)).$$

In other words, the derivative is completely represented geometrically by the tangent cone to the set gph f at the point $(x, f(x))$. By adopting such a geometric characterization as a definition, we can introduce a generalized derivative for an arbitrary set-valued mapping.

© The Author(s), under exclusive license to Springer Nature Switzerland AG 2021
A. L. Dontchev, *Lectures on Variational Analysis*, Applied Mathematical Sciences 205, https://doi.org/10.1007/978-3-030-79911-3_7

Graphical Derivative. *For a mapping $F : I\!R^n \rightrightarrows I\!R^m$ and a pair (x, y) with $y \in F(x)$, the graphical derivative of F at x for y is the mapping $DF(x \,|\, y) : I\!R^n \rightrightarrows I\!R^m$ whose graph is the tangent cone $T_{\text{gph } F}(x, y)$ to* gph F *at (x, y):*

$$v \in DF(x \,|\, y)(u) \iff (u, v) \in T_{\text{gph } F}(x, y).$$

Thus, $v \in DF(x \,|\, y)(u)$ if and only if there exist sequences $u^k \to u$, $v^k \to v$ and $\tau^k \searrow 0$ such that $y + \tau^k v^k \in F(x + \tau^k u^k)$ for all k.

The graphical derivative mappings may no longer be single-valued, but they are positively homogeneous. A mapping H acting between linear spaces is said to be *positively homogeneous* when gph H is a cone, which is equivalent to H satisfying

$$0 \in H(0) \quad \text{and} \quad H(\lambda x) = \lambda H(x) \text{ for } \lambda > 0.$$

Clearly, the inverse of a positively homogeneous mapping is another positively homogeneous mapping. Linear mappings are positively homogeneous as a special case, their graphs being not just cones but linear subspaces. Norm concepts are introduced next in particular for capturing quantitative characteristics.

For any positively homogeneous mapping $H : X \rightrightarrows Y$, where X and Y are Banach spaces, the *outer norm* and the *inner norm* are defined, respectively, by

$$\|H\|^+ = \sup_{\|x\| \le 1} \sup_{y \in H(x)} \|y\| \quad \text{and} \quad \|H\|^- = \sup_{\|x\| \le 1} \inf_{y \in H(x)} \|y\|$$

with the convention $\inf_{y \in \emptyset} \|y\| = \infty$ and $\sup_{y \in \emptyset} \|y\| = -\infty$.

When H is a linear mapping, both $\|H\|^+$ and $\|H\|^-$ reduce to the operator norm $\|H\|$. However, it must be noted that neither $\|H\|^+$ nor $\|H\|^-$ satisfies the conditions in the definition of a norm, inasmuch as set-valued mappings do not even form a vector space.

The inner and outer norms have simple interpretations when $H = A^{-1}$ for a linear and bounded mapping $A : I\!R^n \to I\!R^m$. Let the $m \times n$ matrix for this linear mapping be denoted by A, for simplicity. If $m < n$, we have A surjective (the associated matrix being of rank m) if and only if $\|A^{-1}\|^-$ is finite, this expression being the norm of the right inverse of A: $\|A^{-1}\|^- = \|A^{\mathsf{T}}(AA^{\mathsf{T}})^{-1}\|$. Then $\|A^{-1}\|^+ = \infty$. On the other hand, if $m > n$, we have $\|A^{-1}\|^+ < \infty$ if and only if A is injective (the associated matrix has rank n), and then $\|A^{-1}\|^+ = \|(A^{\mathsf{T}}A)^{-1}A^{\mathsf{T}}\|$ but $\|A^{-1}\|^- = \infty$. For $m = n$, of course, both norms agree with the usual matrix norm $\|A^{-1}\|$, and the finiteness of this quantity is equivalent to nonsingularity of A.

The following norm characterizations of the outer norm and the inner norm of a positively homogeneous mapping H are useful. The inner norm satisfies

$$\|H\|^- = \inf\left\{ \kappa > 0 \,\middle|\, H(x) \cap \kappa I\!B \neq \emptyset \text{ for all } x \in I\!B \right\}. \tag{1}$$

In parallel, for the outer norm we have

$$\|H\|^{+} = \inf\Big\{ \kappa \in (0, \infty) \,\Big|\, y \in H(x) \Rightarrow \|y\| \le \kappa\|x\| \Big\} = \sup_{\|y\|=1} \frac{1}{d(0, H^{-1}(y))}. \quad (2)$$

Since the graphical differentiation comes from an operation on graphs, and the graph of a mapping F can be converted to the graph of its inverse F^{-1} just by interchanging variables, we immediately have the rule that

$$D(F^{-1})(y\,|\,x) = DF(x\,|\,y)^{-1}.$$

Another useful relation is available for sums of a differential function and a set-valued mapping.

Proposition 7.1. *For a function $f : I\!R^n \to I\!R^m$ which is differentiable at x, a set-valued mapping $F : I\!R^n \rightrightarrows I\!R^m$ and any $y \in F(x)$, one has*

$$D(f + F)(x\,|\,f(x) + y) = Df(x) + DF(x\,|\,y).$$

Proof. If $v \in D(f + F)(x\,|\,f(x) + y)(u)$, there exist sequences $\tau^k \searrow 0$ and $u^k \to u$ and $v^k \to v$ such that

$$f(x) + y - f(x + \tau^k u^k) + \tau^k v^k \in F(x + \tau^k u^k) \quad \text{for every } k.$$

By using the definition of the derivative for f we get

$$y + \tau^k(-Df(x)u + v^k) + o(\tau^k) \in F(x + \tau^k u^k).$$

Hence, by the definition of the graphical derivative, $v \in Df(x)u + DF(x\,|\,y)(u)$.

Conversely, if $v - Df(x)u \in DF(x\,|\,y)(u)$ then there exist sequences $\tau^k \searrow 0$, and $u^k \to u$ and $w^k \to v - Df(x)u$ such that $y + \tau^k w^k \in F(x + \tau^k u^k)$. Again, by the differentiability of f,

$$y + f(x) + \tau^k v^k + o(\tau^k) \in (f + F)(x + \tau^k u^k) \quad \text{for} \quad v^k = w^k + Df(x)u^k,$$

which yields $v \in D(f + F)(x\,|\,f(x) + y)(u)$. □

Example. Consider a general constraint system of the form

$$f(x) - D \ni y,$$

for a function $f : I\!R^n \to I\!R^m$, a set $D \subset I\!R^m$ and a parameter vector y, and let $\bar{x}$ be a solution of it for $\bar{y}$ at which f is differentiable. Then for the mapping

$$G : x \mapsto f(x) - D, \quad \text{with} \quad \bar{y} \in G(\bar{x}),$$

one has

$$DG(\bar{x}\,|\,\bar{y})(u) = Df(\bar{x})u - T_D(f(\bar{x}) - \bar{y}).$$

A special but important case of the above example is when $D = I\!R^s_- \times \{0\}^{m-s}$ with $f = (f_1, \ldots, f_m)$, that is,

$$f_i(x) \begin{cases} \le y_i & \text{for } i = 1, \dots, s, \\ = y_i & \text{for } i = s+1, \dots, m. \end{cases}$$

Then a vector $v = (v_1, \dots, v_m)$ is in $DG(x \,|\, y)(u)$ if and only if

$$\nabla f_i(x)u \begin{cases} \le v_i & \text{for } i \in [1, s] \text{ with } f_i(x) = y_i, \\ = v_i & \text{for } i = s+1, \dots, m. \end{cases}$$

Exercise 7.2. *For a function $f : \mathbb{R}^n \to \mathbb{R}^n$ and a convex and closed set $C \subset \mathbb{R}^n$ that is polyhedral, consider the variational inequality*

$$f(x) + N_C(x) \ni y$$

in which N_C is the normal cone mapping and y is a parameter. Let x be a solution at which f is differentiable. Let $v = y - f(x) \in N_C(x)$ and let $K_C(x, v)$ be the corresponding critical cone defined as $K_C(x, v) = T_C(x) \cap [v]^\perp$. Then for the mapping

$$G : x \mapsto f(x) + N_C(x), \quad \textit{with} \quad y \in G(x),$$

one has

$$DG(x \,|\, y)(u) = \nabla f(x)u + N_{K_C(x,v)}(u).$$

Guide. Apply Reduction Lemma given in Lecture 1.

The following theorem presents a graphical derivative criterion for metric regularity:

Theorem 7.3. *For a mapping $F : \mathbb{R}^n \rightrightarrows \mathbb{R}^m$ and a point $(\bar{x}, \bar{y}) \in \operatorname{gph} F$ at which the* gph F *is locally closed, one has*

$$\operatorname{reg}(F; \bar{x} \,|\, \bar{y}) = \limsup_{\substack{(x,y) \to (\bar{x}, \bar{y}) \\ (x,y) \in \operatorname{gph} F}} \|DF(x \,|\, y)^{-1}\|^{-}. \tag{3}$$

Thus, F is metrically regular at $\bar{x}$ for $\bar{y}$ if and only if the right side of (3) *is finite.*

In the case when $m \le n$ and F is a function f which is differentiable on a neighborhood of $\bar{x}$, the representation of the regularity modulus in (3) says that f is metrically regular precisely when the Jacobians $\nabla f(x)$ for x near $\bar{x}$ are of full rank and the inner norms of their inverses $\nabla f(x)^{-1}$ are uniformly bounded. This holds automatically when f is continuously differentiable around $\bar{x}$ with $\nabla f(\bar{x})$ of full rank, which is the same as surjective. When $m = n$ this becomes nonsingularity and we come to the classical inverse function theorem.

Exercise 7.4. *Consider the sum $f + F$ of a function $f : \mathbb{R}^n \to \mathbb{R}^m$ and a set-valued mapping $F : \mathbb{R}^n \rightrightarrows \mathbb{R}^m$, let $0 \in f(\bar{x}) + F(\bar{x})$, let f be differentiable around $\bar{x}$, and let F have locally closed graph at $\bar{x}$ for $-f(\bar{x})$. Show that the mapping $f + F$ is metrically regular at $\bar{x}$ for* 0 *provided that*

$$\limsup_{\substack{(x,y)\to(\bar{x},0)\\ y\in f(x)+F(x)}} \|(\nabla f(x) + DF(x\,|\,y - f(x)))^{-1}\|^{-} < \infty.$$

Guide. Apply Theorem 7.3 combined with the sum rule in Proposition 7.1. □

In Lecture 1 we introduced the concept of a normal cone $N_C(x)$ to a closed convex set. A general normal cone to a possibly nonconvex set C is now introduced.

Regular and General Normal Cones. *For a set $C \subset I\!R^n$ and a point $x \in C$ at which C is locally closed, a vector v is said to be a regular normal if $\langle v, x'-x\rangle \le o(\|x'-x\|)$ for $x' \in C$. The set of all such vectors v is called the regular normal cone to C at x and is denoted by $\hat{N}_C(x)$. A vector v is said to be a general normal to C at x if there are sequences $\{x^k\}$ and $\{v^k\}$ with $x^k \in C$, such that*

$$x^k \to x \text{ and } v^k \to v \text{ with } v^k \in \hat{N}_C(x^k).$$

The set of all such vectors v is called the general normal cone to C at x and is denoted by $\bar{N}_C(x)$. For $x \notin C$, $\bar{N}_C(x)$ is the empty set. When the set C is closed and convex, the general normal cone $\bar{N}_C(x)$ becomes the cone $N_C(x)$ introduced in Lecture 1.

Another concept of generalized differentiation that can be used to characterize metric regularity is the so-called *coderivative*.

Coderivative. *For a mapping $F : I\!R^n \rightrightarrows I\!R^m$ and a pair $(x, y) \in \operatorname{gph} F$ at which $\operatorname{gph} F$ is locally closed, the coderivative of F at x for y is the mapping $D^*F(x\,|\,y) : I\!R^m \rightrightarrows I\!R^n$ defined by*

$$w \in D^*F(x\,|\,y)(z) \iff (w, -z) \in \bar{N}_{\operatorname{gph} F}(x, y).$$

In the case where F is single-valued, thus reducing to a function $f : I\!R^n \to I\!R^m$, suppose f is strictly differentiable at x; then for $y = f(x)$, the graphical derivative $Df(x\,|\,y)$ is of course the linear mapping $Df(x)$ from $I\!R^n$ to $I\!R^m$ with matrix $\nabla f(x)$. In contrast, the coderivative $D^*f(x\,|\,y)$ comes out as the adjoint linear mapping $Df(x)^*$ from $I\!R^m$ to $I\!R^n$ with matrix $\nabla f(x)^{\mathsf{T}}$.

A coderivative criterion for metric regularity is presented next.

Theorem 7.5. *For a mapping $F : I\!R^n \rightrightarrows I\!R^m$ and a pair $(\bar{x}, \bar{y}) \in \operatorname{gph} F$ at which $\operatorname{gph} F$ is locally closed, one has*

$$\operatorname{reg}(F; \bar{x}\,|\,\bar{y}) = \|D^*F(\bar{x}\,|\,\bar{y})^{-1}\|^{+}. \tag{4}$$

Thus, F is metrically regular if and only if the right side of (4) *is finite.*

If F is single-valued, that is, a function $f : I\!R^n \to I\!R^m$ which is strictly differentiable at $\bar{x}$, then the coderivative criterion means that the adjoint to the derivative mapping $Df(\bar{x})$ is injective, that is, $\ker \nabla f(x)^{\mathsf{T}} = \{0\}$. This is equivalent to surjectivity of $Df(\bar{x})$ which, as we know, is equivalent to metric regularity of f at $\bar{x}$ for $f(\bar{x})$.

Theorem 7.5 follows from the following result, whose proof is given at the end of the lecture:

Theorem 7.6. *Let $F : I\!R^n \rightrightarrows I\!R^m$ be a set-valued map, let $\bar{y} \in F(\bar{x})$, and assume that* gph F *is locally closed at $(\bar{x}, \bar{y})$. Then*

$$\limsup_{\substack{(x,y)\to(\bar{x},\bar{y}),\\ (x,y)\in \operatorname{gph} F}} \|DF(x\,|\,y)^{-1}\|^{-} = \|D^*F(\bar{x}\,|\,\bar{y})^{-1}\|^{+}. \tag{5}$$

Proof of Theorem 7.5. If F is metrically regular at $\bar{x}$ for $\bar{y}$, then the expression on the left side of (5) is finite and equal to the regularity modulus. Then the expression on the right of (5) is finite as well and equal to the regularity modulus, which implies (4). Conversely, if the expression on the right side of (4) is finite, then so is the expression on the left side of (5), which, by invoking again Theorem 7.6, equals to the regularity modulus of F at $\bar{x}$ for $\bar{y}$. Hence F is metrically regular at $\bar{x}$ for $\bar{y}$. □

We end this lecture with two exercises. This first is a sum rule for the coderivative, while the second one specializes the coderivative criterion for generalized equations.

Exercise 7.7. *For a function $f : I\!R^n \to I\!R^m$ which is strictly differentiable at $\bar{x}$ and a mapping $F : I\!R^n \rightrightarrows I\!R^m$ with $\bar{y} \in F(\bar{x})$, prove that*

$$D^*(f + F)(\bar{x}\,|\,f(\bar{x}) + \bar{y})(u) = D^*f(\bar{x})u + D^*F(\bar{x}\,|\,\bar{y})(u) \quad \textit{for all} \quad u \in I\!R^m.$$

Exercise 7.8. *For the solution mapping $S(p) = \{\, x \,|\, f(p, x) + F(x) \ni 0 \,\}$ and a pair $(\bar{p}, \bar{x})$ with $\bar{x} \in S(\bar{p})$, suppose that f is differentiable in a neighborhood of $(\bar{p}, \bar{x})$,* gph F *is locally closed at $(\bar{x}, -f(\bar{p}, \bar{x}))$, and*

$$\|(D_x^* f(p, x) + D^*F(x\,|\,y - f(p, x)))^{-1}\|^{+} \le \lambda < \infty.$$

Then S has the Aubin property at $\bar{p}$ for $\bar{x}$ with

$$\operatorname{lip}(S; \bar{p}\,|\,\bar{x}) \le \lambda\, \widehat{\operatorname{lip}}_p(f; \bar{p}\,|\,\bar{x}).$$

Proofs

The proof of Theorem 7.3 presented in further lines is based on a solution mapping estimate for the parameterized inclusion

$$G(p, x) \ni 0 \quad \text{for a mapping} \quad G : I\!R^d \times I\!R^n \rightrightarrows I\!R^m. \tag{6}$$

It makes use of the *partial* graphical derivative of $G(p, x)$ with respect to x, which is defined as the graphical derivative of the mapping $x \mapsto G(p, x)$ with p fixed and denoted by $D_xG(p, x\,|\,y)$. Of course, $D_pG(p, x\,|\,y)$ has a similar meaning.

Theorem 7.9. *For the inclusion* (6) *and its solution mapping*

$$p \mapsto S(p) = \{x \mid G(p, x) \ni 0\},$$

let $\bar{x} \in S(\bar{p})$*, so that* $(\bar{p}, \bar{x}, 0) \in \operatorname{gph} G$*. Suppose that* gph G *is locally closed at* $(\bar{p}, \bar{x}, 0)$ *and that the distance function* $p \mapsto d(0, G(p, \bar{x}))$ *is upper semicontinuous at* $\bar{p}$*. Then for every* $c \in (0, \infty)$ *satisfying*

$$\limsup_{\substack{(p,x,y)\to(\bar{p},\bar{x},0)\\ (p,x,y)\in \operatorname{gph} G}} \|D_x G(p, x \,|\, y)^{-1}\|^{-} < c \tag{7}$$

there are neighborhoods V *of* $\bar{p}$ *and* U *of* $\bar{x}$ *such that*

$$d(x, S(p)) \le c\, d(0, G(p, x)) \quad \textit{for } x \in U \textit{ and } p \in V. \tag{8}$$

Proof. Let c satisfy (7). Then there exists $\eta > 0$ such that

$$\begin{cases} \text{for every } (p, x, y) \in \operatorname{gph} G \text{ with } \|p - \bar{p}\| + \max\{\|x - \bar{x}\|, c\|y\|\} \le 2\eta, \\ \text{and for every } v \in I\!\!R^m, \text{ there exists } u \in D_x G(p, x \,|\, y)^{-1}(v) \\ \text{such that } \|u\| \le c\|v\|. \end{cases} \tag{9}$$

We can always choose η smaller so that the intersection

$$\operatorname{gph} G \bigcap \big\{ (p, x, y) \,\big|\, \|p - \bar{p}\| + \max\{\|x - \bar{x}\|, c\|y\|\} \le 2\eta \big\} \quad \text{is closed.} \tag{10}$$

The next part of the proof is presented as a lemma.

Lemma 7.10. *For* c *and* η *as above, let* $\varepsilon > 0$ *and* $s > 0$ *be such that*

$$c\varepsilon < 1 \quad \textit{and} \quad s < \varepsilon\eta \tag{11}$$

and let $(p, \omega, v) \in \operatorname{gph} G$ *satisfy*

$$\|p - \bar{p}\| + \max\{\|\omega - \bar{x}\|, c\|v\|\} \le \eta. \tag{12}$$

Then for every $y' \in I\!\!B_s(v)$ *there exists* $\hat{x}$ *with* $y' \in G(p, \hat{x})$ *such that*

$$\|\hat{x} - \omega\| \le \frac{1}{\varepsilon}\|y' - v\|. \tag{13}$$

In the proof of Lemma 7.10, we apply an important theorem in variational analysis called the Ekeland variational principle, which we state next in the following form:

Ekeland Variational Principle. *Let* (X, ρ) *be a complete metric space and let* $f : X \to (-\infty, \infty]$ *be a lower semicontinuous function on* X *which is bounded from below. Let* $\bar{u} \in \operatorname{dom} f$*. Then for every* $\delta > 0$ *there exists* u_δ *such that*

$$f(u_\delta) + \delta\rho(u_\delta, \bar{u}) \le f(\bar{u}),$$

and

$$f(u_\delta) < f(u) + \delta\rho(u, u_\delta) \quad \textit{for every} \;\; u \in X,\; u \neq u_\delta.$$

Proof of Lemma 7.10. On the product space $Z := I\!R^n \times I\!R^m$ we introduce the norm

$$\|(x, y)\| := \max\{\|x\|, c\|y\|\},$$

which is equivalent to the Euclidean norm. Pick ε, s and $(p, \omega, v) \in \operatorname{gph} G$ as required in (11) and (12) and let $y' \in I\!B_s(v)$. By (10) the set

$$E_p := \{\,(x, y) \,|\, (p, x, y) \in \operatorname{gph} G, \|p - \bar{p}\| + \|(x, y) - (\bar{x}, 0)\| \leq 2\eta\} \subset I\!R^n \times I\!R^m$$

is closed, hence, equipped with the metric induced by the norm in question, it is a complete metric space. The function $V_p : E_p \to I\!R$ defined by

$$V_p : (x, y) \mapsto \|y' - y\| \qquad \text{for} \;\; (x, y) \in E_p \tag{14}$$

is continuous on its domain E_p. Also, $(\omega, v) \in \operatorname{dom} V_p$. We apply the Ekeland variational principle to the function V_p with $\bar{u} = (\omega, v)$ and the chosen ε, to obtain the existence of $(\hat{x}, \hat{y}) \in E_p$ such that

$$V_p(\hat{x}, \hat{y}) + \varepsilon\|(\omega, v) - (\hat{x}, \hat{y})\| \leq V_p(\omega, v) \tag{15}$$

and

$$V_p(\hat{x}, \hat{y}) \leq V_p(x, y) + \varepsilon\|(x, y) - (\hat{x}, \hat{y})\| \qquad \text{for every} \;\; (x, y) \in E_p. \tag{16}$$

With V_p as in (14), the inequalities (15) and (16) come down to

$$\|y' - \hat{y}\| + \varepsilon\|(\omega, v) - (\hat{x}, \hat{y})\| \leq \|y' - v\| \tag{17}$$

and

$$\|y' - \hat{y}\| \leq \|y' - y\| + \varepsilon\|(x, y) - (\hat{x}, \hat{y})\| \qquad \text{for every} \;\; (x, y) \in E_p. \tag{18}$$

Through (17) we obtain in particular that

$$\|(\omega, v) - (\hat{x}, \hat{y})\| \leq \frac{1}{\varepsilon}\|y' - v\|. \tag{19}$$

Since $y' \in I\!B_s(v)$, we then have

$$\|(\omega, v) - (\hat{x}, \hat{y})\| \leq \frac{s}{\varepsilon}$$

and consequently, from the choice of (p, ω, v) in (12) and s in (11),

$$\begin{aligned} &\|p - \bar{p}\| + \|(\hat{x}, \hat{y}) - (\bar{x}, 0)\| \\ &\qquad \leq \|p - \bar{p}\| + \|(\omega, v) - (\bar{x}, 0)\| + \|(\omega, v) - (\hat{x}, \hat{y})\| \leq \eta + \frac{s}{\varepsilon} < 2\eta. \end{aligned} \tag{20}$$

Thus, $(p, \hat{x}, \hat{y})$ satisfies the condition in (9), so there exists $u \in I\!R^n$ for which

$$y' - \hat{y} \in D_x G(p, \hat{x} \,|\, \hat{y})(u) \quad \text{and} \quad \|u\| \le c\|y' - \hat{y}\|. \tag{21}$$

By the definition of the partial graphical derivative, there exist sequences $\tau^k \searrow 0$, $u^k \to u$, and $v^k \to y' - \hat{y}$ such that

$$\hat{y} + \tau^k v^k \in G(p, \hat{x} + \tau^k u^k) \quad \text{for all } k.$$

Also, from (20) we know that, for sufficiently large k,

$$\|p - \bar{p}\| + \|(\hat{x} + \tau^k u^k, \hat{y} + \tau^k v^k) - (\bar{x}, 0)\| \le 2\eta,$$

implying $(\hat{x} + \tau^k u^k, \hat{y} + \tau^k v^k) \in E_p$. If we now plug the point $(\hat{x} + \tau^k u^k, \hat{y} + \tau^k v^k)$ into (18) in place of (x, y), we get

$$\|y' - \hat{y}\| \le \|y' - (\hat{y} + \tau^k v^k)\| + \varepsilon\|(\hat{x} + \tau^k u^k, \hat{y} + \tau^k v^k) - (\hat{x}, \hat{y})\|.$$

This gives us

$$\|y' - \hat{y}\| \le (1 - \tau^k)\|y' - \hat{y}\| + \tau^k\|v^k - (y' - \hat{y})\| + \varepsilon\tau^k\|(u^k, v^k)\|,$$

that is,

$$\|y' - \hat{y}\| \le \|v^k - (y' - \hat{y})\| + \varepsilon\|(u^k, v^k)\|.$$

Passing to the limit with $k \to \infty$ leads to $\|y' - \hat{y}\| \le \varepsilon\|(u, y' - \hat{y})\|$ and then, taking into account the second relation in (21), we conclude that $\|y' - \hat{y}\| \le \varepsilon c\|y' - \hat{y}\|$. Since $\varepsilon c < 1$ by (9), the only possibility here is that $y' = \hat{y}$. But then $y' \in G(p, \hat{x})$ and (19) yields (13). This proves the lemma. □

We continue with the proof of Theorem 7.9. Let $\tau = \eta/(4c)$. Since the function $p \to d(0, G(p, \bar{x}))$ is upper semicontinuous at $\bar{p}$, there exists a positive $\delta \le c\tau$ such that $d(0, G(p, \bar{x})) \le \tau/2$ for all p with $\|p - \bar{p}\| < \delta$. Set $V := I\!B_\delta(\bar{p})$, $U := I\!B_{c\tau}(\bar{x})$ and pick any $p \in V$ and $x \in U$. We can find y such that $y \in G(p, \bar{x})$ with $\|y\| \le d(0, G(p, \bar{x})) + \tau/3 < \tau$. Note that

$$\|p - \bar{p}\| + \|(\bar{x}, y) - (\bar{x}, 0)\| = \|p - \bar{p}\| + c\|y\| \le \delta + c\tau \le \eta. \tag{22}$$

Choose $\varepsilon > 0$ such that $1/2 < \varepsilon c < 1$ and let $s = \varepsilon\eta/2$. Then $s > \tau$. We apply Lemma 7.8 with the indicated ε and s, and with $(p, \omega, v) = (p, \bar{x}, y)$ which, as seen in (22), satisfies (12), and with $y' = 0$, since $0 \in I\!B_s(y)$. Thus, there exists $\hat{x}$ such that $0 \in G(p, \hat{x})$, that is, $\hat{x} \in S(p)$, and also, from (13), $\|\hat{x} - \bar{x}\| \le \|y|\varepsilon$. Therefore, in view of the choice of y, we have $\hat{x} \in I\!B_{\tau/\varepsilon}(\bar{x})$. We now consider two cases.

Case 1. $d(0, G(p, x)) \ge 2\tau$. We just proved that there exists $\hat{x} \in S(p)$ with $\hat{x} \in I\!B_{\tau/\varepsilon}(\bar{x})$; then

$$\begin{aligned} d(x, S(p)) &\le d(\bar{x}, S(p)) + \|x - \bar{x}\| \\ &\le \|\bar{x} - \hat{x}\| + \|x - \bar{x}\| \le \frac{\tau}{\varepsilon} + c\tau \le \frac{2\tau}{\varepsilon} \le \frac{1}{\varepsilon} d(0, G(p, x)). \end{aligned} \tag{23}$$

CASE 2. $d(0, G(p, x)) < 2\tau$. In this case, for any y with $\|y\| \le 2\tau$ we have

$$\|p - \bar{p}\| + \max\{\|x - \bar{x}\|, c\|y\|\} \le \delta + \max\{c\tau, 2c\tau\} \le 3c\tau \le \eta$$

and then, by (10), the nonempty set $G(p, x) \cap 2\tau I\!B$ is closed. Hence, there exists $\tilde{y} \in G(p, x)$ such that $\|\tilde{y}\| = d(0, G(p, x)) < 2\tau$ and therefore

$$c\|\tilde{y}\| < 2c\tau = \frac{\eta}{2}\,.$$

We conclude that the point $(p, x, \tilde{y}) \in \operatorname{gph} G$ satisfies

$$\|p - \bar{p}\| + \max\{\|x - \bar{x}\|, c\|\tilde{y}\|\} \le \delta + \max\{c\tau, \eta/2\} \le \eta.$$

Thus, the assumptions of Lemma 7.8 hold for $(p, \omega, \nu) = (p, x, \tilde{y})$, $s = 2\tau$, and $y' = 0$. Hence there exists $\tilde{x} \in S(p)$ such that

$$\|\tilde{x} - x\| \le \frac{1}{\varepsilon}\|\tilde{y}\|.$$

Then, by the choice of $\tilde{y}$,

$$d(x, S(p)) \le \|x - \tilde{x}\| \le \frac{1}{\varepsilon}\|\tilde{y}\| = \frac{1}{\varepsilon} d(0, G(p, x)).$$

Hence, by (23), for both cases 1 and 2, and therefore for any p in V and $x \in U$, we have

$$d(x, S(p)) \le \frac{1}{\varepsilon} d(0, G(p, x)).$$

Since U and V do not depend on ε, and $1/\varepsilon$ can be arbitrarily close to c, this gives us (8). □

Proof of Theorem 7.3. For short, let d_{DF} denote the right side of (3). We will start by showing that $\operatorname{reg}(F; \bar{x}\,|\,\bar{y}) \le d_{DF}$. If $d_{DF} = \infty$ there is nothing to prove. Let $d_{DF} < c < \infty$. Applying Theorem 7.7 to $G(p, x) = F(x) - p$ and this c, letting y take the place of p, we have $S(y) = F^{-1}(y)$ and $d(0, G(y, x)) = d(y, F(x))$. Condition (8) becomes the definition of metric regularity of F at $\bar{x}$ for $\bar{y} = \bar{p}$, and therefore $\operatorname{reg}(F; \bar{x}\,|\,\bar{y}) \le c$. Since c can be arbitrarily close to d_{DF}, we conclude that $\operatorname{reg}(F; \bar{x}\,|\,\bar{y}) \le d_{DF}$.

We turn now to demonstrating the opposite inequality,

$$\operatorname{reg}(F; \bar{x}\,|\,\bar{y}) \ge d_{DF}. \tag{24}$$

If $\operatorname{reg}(F; \bar{x}\,|\,\bar{y}) = \infty$ we are done. Suppose therefore that F is metrically regular at $\bar{x}$ for $\bar{y}$ with respect to a constant κ and neighborhoods U for $\bar{x}$ and V for $\bar{y}$. Then

$$d(x', F^{-1}(y)) \le \kappa\|y - y'\| \quad \text{whenever } (x', y') \in \operatorname{gph} F,\ x' \in U,\ y \in V. \tag{25}$$

We know from Proposition 4.1 that V can be chosen so small that $F^{-1}(y) \cap U \neq \emptyset$ for every $y \in V$. Pick any $y' \in V$ and $x' \in F^{-1}(y') \cap U$, and let $v \in \mathbb{B}$. Take a sequence $\tau^k \searrow 0$ such that $y^k := y' + \tau^k v \in V$ for all k. By (25) and the local closedness of gph F at $(\bar{x}, \bar{y})$, there exists $x^k \in F^{-1}(y' + \tau^k v)$ such that

$$\|x' - x^k\| = d(x', F^{-1}(y^k)) \leq \kappa \|y^k - y'\| = \kappa \tau^k \|v\|.$$

For $u^k := (x^k - x')/\tau^k$ we obtain

$$\|u^k\| \leq \kappa \|v\|. \tag{26}$$

Thus, u^k is bounded, so $u^{k_i} \to u$ for a subsequence $k_i \to \infty$. Since $(x^{k_i}, y' + \tau^{k_i} v) \in \text{gph } F$, we obtain $(u, v) \in T_{\text{gph } F}(x', y')$. Hence, by the definition of the graphical derivative, we have $u \in DF^{-1}(y' \mid x')(v) = DF(x' \mid y')^{-1}(v)$. The bound (26) guarantees that

$$\|DF(x \mid y)^{-1}\|^{-} \leq \kappa.$$

Since $(y, x) \in \text{gph } S$ is arbitrarily chosen near $(\bar{x}, \bar{y})$, and κ is independent of this choice, we conclude that (24) holds and hence we have (3). □

In the proof of Theorem 7.6, we employ the following lemma:

Lemma 7.11. *Let C be a convex and compact set in $\mathbb{R}^d$, $K \subset \mathbb{R}^d$ be a closed set and $\bar{x} \in K$. Then $C \cap T_K(x) \neq \emptyset$ for all $x \in K$ near $\bar{x}$ if and only if $C \cap \text{clco}\, T_K(x) \neq \emptyset$ for all $x \in K$ near $\bar{x}$.*

Proof. Clearly, $C \cap T_K(x) \neq \emptyset$ implies $C \cap \text{clco}\, T_K(x) \neq \emptyset$. Assume that there exists an open neighborhood U of $\bar{x}$ such that $C \cap \text{clco}\, T_K(x) \neq \emptyset$ for all $x \in K \cap U$. Let $\varepsilon > 0$ be such that $\mathbb{B}_\varepsilon(\bar{x}) \subset U$. Take any $x \in \mathbb{B}_{\varepsilon/3}(\bar{x})$and let v be a projection of x on K. Then $\|v - x\| \leq \|\bar{x} - x\| \leq \varepsilon/3$ and hence

$$\|v - \bar{x}\| \leq \|v - x\| + \|x - \bar{x}\| \leq \varepsilon/3 + \varepsilon/3 < \varepsilon.$$

Thus, there exists an open neighborhood W of $\bar{x}$ such that any metric projection of a point $x \in W$ on K belongs to $K \cap U$.

Fix $x \in K \cap W$. For all $t \geq 0$ define $\varphi(t) := \min\{\|u - v\| \mid u \in x + tC, v \in K\}$. The function φ is Lipschitz continuous. Indeed, for any $t_i \geq 0$, $i = 1, 2$ there exist $c_i \in C$ and $k_i \in K$ such that $\varphi(t_i) = \|x + t_i c_i - k_i\|$, $i = 1, 2$. Then

$$\begin{aligned} \varphi(t_1) - \varphi(t_2) &= \|x + t_1 c_1 - k_1\| - \|x + t_2 c_2 - k_2\| \\ &\leq \|x + t_1 c_2 - k_2\| - \|x + t_2 c_2 - k_2\| \leq \|c_2\| \|t_1 - t_2\|. \end{aligned}$$

This implies that the derivative φ' exists almost everywhere and $\varphi(s) = \varphi(t) + \int_t^s \varphi'(\tau) d\tau$ for all $s \geq t \geq 0$. We will prove next that

$$\varphi(t) = 0 \text{ for all sufficiently small } t > 0. \tag{27}$$

If this holds, then for every small $t > 0$ there exists $v_t \in C$ such that $x + tv_t \in K$. Consider sequences $t_k \searrow 0$ and $v_{t_k} \in C$ such that v_{t_k} converges to some v. Then $v \in T_K(x) \cap C$ and since $x \in K \cap W$ is arbitrary, we arrive at the claim of the lemma.

To prove (27), let $\gamma > 0$ be such that $x + [0, \gamma]C \subset W$. Assume that there exists $t_0 \in (0, \gamma]$ such that $\varphi(t_0) > 0$. Define $\bar{t} = \max\{t \mid \varphi(t) = 0 \text{ and } 0 \leq t < t_0\}$. Let $t \in (\bar{t}, t_0]$ be such that $\varphi'(t)$ exists. Then for some $v_t \in C$ and $x_t \in K$ we have $\varphi(t) = \|x + tv_t - x_t\| > 0$. Since x_t is a projection of $x + tv_t$ on K, by the observation in the beginning of the proof we have $x_t \in K \cap U$. By assumption, there exists $w_t \in \operatorname{clco} T_K(x_t)$ such that $w_t \in C$. Then, for any $h > 0$ sufficiently small,

$$x + tv_t + hw_t = x + (t + h)\left(\frac{t}{t + h}v_t + \frac{h}{t + h}w_t\right) \in x + (t + h)C \subset W$$

because the set C is assumed convex. Thus

$$\varphi(t + h) - \varphi(t) \leq \|x + tv_t + hw_t - x_t\| - \|x + tv_t - x_t\|.$$

Dividing both sides of this inequality by $h > 0$ and passing to the limit when $h \to 0_+$, we get

$$\varphi'(t) \leq \left\langle \frac{x + tv_t - x_t}{\|x + tv_t - x_t\|}, w_t \right\rangle.$$

Recall that x_t is a projection of $x + tv_t$ on K and also the elementary fact that in this case $x + tv_t - x_t \in \hat{N}_K(x_t)$. Since $w_t \in \operatorname{clco} T_K(x_t)$, we obtain from the inequality above that $\varphi'(t) \leq 0$. Having in mind that t is any point of differentiability of φ in $(\bar{t}, t_0)$, we get $\varphi(t_0) \leq \varphi(\bar{t}) = 0$. This contradicts the choice of t_0 according to which $\varphi(t_0) > 0$. Hence (27) holds and the lemma is proved. □

In the proof of Theorem 7.6, we use the so-called *separation theorem*:

Separation of a Convex Set from a Point. *Let C be a nonempty, closed, convex subset of $I\!R^n$, and let $x_0 \in X$. Then $x_0 \notin C$ if and only if there exists $u \in I\!R^n$ such that*

$$u^{\mathsf{T}} x_0 > \sup_{x \in C} u^{\mathsf{T}} x.$$

Essentially, this says geometrically that a closed convex set is the intersection of all the "closed half-spaces" that include it. It is a corollary of a more general result concerning separation by a hyperplane of two convex sets that do not intersect each other (Fig. 7.1).

Proof of Theorem 7.6. Since the graphical derivative and the coderivative are defined only locally around $(\bar{x}, \bar{y})$, we can assume without loss of generality that the graph of the mapping F is closed. We will show first that

$$\limsup_{\substack{(x,y)\to(\bar{x},\bar{y}),\\ (x,y)\in \operatorname{gph} F}} \|DF(x \mid y)^{-1}\|^{-} \geq \|D^*F(\bar{x} \mid \bar{y})^{-1}\|^{+}. \tag{28}$$

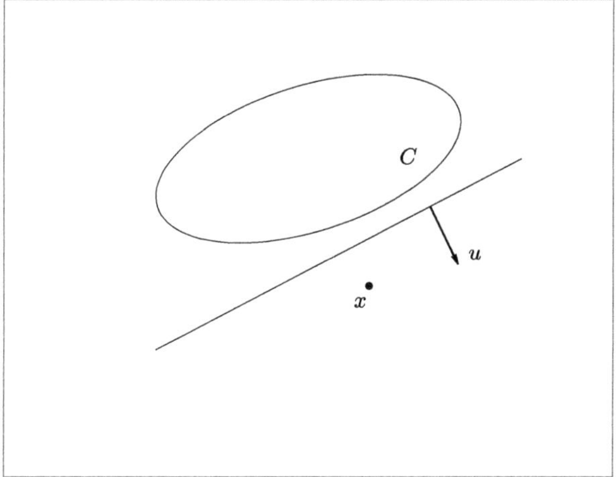

Fig. 7.1: Separation of a convex set from a point

If the left side of (28) equals ∞, there is nothing to prove. Suppose that a constant c satisfies

$$\limsup_{\substack{(x,y)\to(\bar{x},\bar{y}),\\ (x,y)\in \operatorname{gph} F}} \|DF(x\,|\,y)^{-1}\|^{-} < c.$$

From the property (2) of the outer norm there exists $\delta > 0$ such that for all $(x, y) \in \operatorname{gph} F \cap (I\!B_\delta(\bar{x}) \times I\!B_\delta(\bar{y}))$ and for every $v \in I\!B$ there exists $u \in DF(x\,|\,y)^{-1}(v)$ such that $|u| < c$. Also, note that

$$(u, v) \in T_{\operatorname{gph} F}(x, y) \subset \operatorname{clco} T_{\operatorname{gph} F}(x, y) = T^{**}_{\operatorname{gph} F}(x, y).$$

Fix $(x, y) \in \operatorname{gph} F \cap (I\!B_\delta(\bar{x}) \times I\!B_\delta(\bar{y}))$ and let $v \in I\!B \subset I\!R^m$. Then there exists u with $(u, v) \in T^{**}_{\operatorname{gph} F}(x, y)$ such that $u = cw$ for some $w \in I\!B$. Let $(p, q) \in \hat{N}_{\operatorname{gph} F}(x, y) = T^{*}_{\operatorname{gph} F}(x, y)$. From the inequality $\langle u, p\rangle + \langle v, q\rangle \leq 0$ we get

$$c \min_{w\in I\!B} \langle w, p\rangle + \langle v, q\rangle \leq 0 \quad \text{which yields} \quad -c\|p\| + \langle v, q\rangle \leq 0.$$

Since v is arbitrarily chosen in $I\!B$, we conclude that

$$\|q\| \leq c\|p\| \;\text{ whenever }\; (p, q) \in \hat{N}_{\operatorname{gph} F}(x, y). \tag{29}$$

Now, let $(-p, q) \in N_{\operatorname{gph} F}(\bar{x}, \bar{y})$; then there exist sequences $(x_k, y_k) \in \operatorname{gph} F$, $(x_k, y_k) \to (\bar{x}, \bar{y})$ and $(-p_k, q_k) \in \hat{N}_{\operatorname{gph} F}(x_k, y_k)$ such that $(-p_k, q_k) \to (-p, q)$. But then, from (29), $\|q_k\| \leq c\|p_k\|$ and in the limit $\|q\| \leq c\|p\|$. Thus, $\|q\| \leq c\|p\|$ whenever $(-p, q) \in N_{\operatorname{gph} F}(\bar{x}, \bar{y})$ and therefore we have $\|q\| \leq c\|p\|$ whenever $(q, -p) \in N_{\operatorname{gph} F^{-1}}(\bar{y}, \bar{x})$. By the definition of the coderivative,

$$\|q\| \le c\|p\| \text{ whenever } q \in D^*F(\bar{x}\,|\,\bar{y})^{-1}(p).$$

This, together with the property of the inner norm given in (1), implies that $c \ge \|D^*F(x,y)^{-1}\|^+$ and we obtain (28) since c is arbitrary.

For the converse inequality, it is enough to consider the case when there is a constant c such that

$$\|D^*F(\bar{x}\,|\,\bar{y})^{-1}\|^+ < c. \tag{30}$$

We first show there exists $\delta > 0$ such that for $(x,y) \in \text{gph } F \cap (I\!B_\delta(\bar{x}) \times I\!B_\delta(\bar{y}))$ we have that

$$(0,v) \in \hat{N}_{\text{gph } F}(x,y) \implies v = 0. \tag{31}$$

On the contrary, assume that there exist sequences $(x_k, y_k) \in \text{gph } F$ with $(x_k, y_k) \to (\bar{x}, \bar{y})$ and $v_k \in I\!R^m$ with $\|v_k\| = 1$ such that $(0, v_k) \in \hat{N}_{\text{gph } F}(x_k, y_k)$ for all k. There is $v \ne 0$ such that $(0,v) \in N_{\text{gph } F}(\bar{x}, \bar{y})$, that is $v \in D^*F(\bar{x}\,|\,\bar{y})^{-1}(0)$. Taking into account (2), this contradicts (30).

We will now prove a statement more general than (31); namely, there exists $\delta > 0$ such that for any $(x,y) \in \text{gph } F \cap (I\!B_\delta(\bar{x}) \times I\!B_\delta(\bar{y}))$ we have

$$(v, -u) \in \hat{N}_{\text{gph } F^{-1}}(y,x) \implies \|v\| \le c\|u\|. \tag{32}$$

On the contrary, assume that there exists a sequence $(y_k, x_k) \to (\bar{y}, \bar{x})$ such that for each k we can find $(v_k, -u_k) \in \hat{N}_{\text{gph } F^{-1}}(y_k, x_k)$ satisfying $\|v_k\| > c\|u_k\|$. If $u_k = 0$ for some k, then from (31) we get $v_k = 0$, a contradiction. Thus, without loss of generality we assume that $\|u_k\| = 1$. Let v_k be unbounded and let w be a cluster point of $\frac{1}{\|v_k\|}v_k$; then $\|w\| = 1$. Since $(\frac{1}{\|v_k\|}v_k, -\frac{1}{\|v_k\|}u_k) \in \hat{N}_{\text{gph } F^{-1}}(y_k, x_k)$, passing to the limit we get $(w, 0) \in N_{\text{gph } F^{-1}}(\bar{y}, \bar{x})$ which contradicts (30) because of (2). Further, if v_k is bounded, then $(v_k, u_k) \to (v, u)$ for a subsequence, where $\|u\| = 1$, $(v, -u) \in N_{\text{gph } F^{-1}}(\bar{y}, \bar{x})$, and $\|v\| \ge c$. This again contradicts (30). Thus, (32) holds for all $(y,x) \in \text{gph } F^{-1}$ close to $(\bar{y}, \bar{x})$.

Let $\delta > 0$ be such that (32) is satisfied for any $(x,y) \in \text{gph } F \cap (I\!B_\delta(\bar{x}) \times I\!B_\delta(\bar{y}))$. Pick such (x,y). We will show that

$$(cI\!B \times \{w\}) \cap T^{**}_{\text{gph } F}(x,y) \ne \emptyset \quad \text{for every } w \in I\!B. \tag{33}$$

On the contrary, assume that there exists $w \in I\!B$ such that $(cI\!B \times \{w\}) \cap T^{**}_{\text{gph } F}(x,y) = \emptyset$. Then, by the theorem on separation of a convex set from a point stated before the proof, there exists a nonzero $(p,q) \in T^*_{\text{gph } F}(x,y) = \hat{N}_{\text{gph } F}(x,y)$ such that

$$\min_{u \in I\!B} \langle p, cu \rangle + \langle q, w \rangle > 0.$$

Indeed, it is sufficient to choose a hyperplane through the origin that separates the sets $(cI\!B \times \{w\})$ and $T^{**}_{\text{gph } F}(x,y)$; then take the normal (p,q) pointing toward the half-space containing $(cI\!B \times \{w\})$. If $p = 0$, then $q \ne 0$ and then $(q, 0) \in \hat{N}_{\text{gph } F^{-1}}(y,x)$ in contradiction with (32). Hence, $p \ne 0$. Without loss of generality, let $\|p\| = 1$. Then $(q,p) \in \hat{N}_{\text{gph } F^{-1}}(y,x)$ and

$$\langle q, w\rangle > \max_{u\in \mathbb{B}}\langle p, cu\rangle = c\|p\| = c. \tag{34}$$

By (2) and (32), $|q| \leq c$ and since $w \in \mathbb{B}$, this contradicts (34). Thus, (33) is satisfied.

By Lemma 7.9, for all $(x, y) \in \operatorname{gph} F$ sufficiently close to $(\bar{x}, \bar{y})$, we have that (33) holds when the set $T^{**}_{\operatorname{gph} F}(x, y) = \operatorname{clco} T_{\operatorname{gph} F}(x, y)$ is replaced with $T_{\operatorname{gph} F}(x, y)$. This means that for any $w \in \mathbb{B}$ there exists $u \in DF(x\,|\,y)^{-1}(w)$ such that $\|u\| \leq c$. But then $c \geq \|DF(x\,|\,y)^{-1}\|^{-}$ for all $(x, y) \in \operatorname{gph} F$ sufficiently close to $(\bar{x}, \bar{y})$. This combined with the arbitrariness of c in (30) implies the inequality opposite to (26) and hence the proof of the theorem if complete. $\square$

Lecture 8
Strong Regularity

We begin this lecture with a basic theorem in analysis: the classical inverse function theorem. Recall that, for a function $f : X \to Y$, where X and Y are nonempty sets, we define the inverse of f as the mapping

$$Y \ni y \mapsto f^{-1}(y) := \{x \in X \mid f(x) = y\}.$$

Thus, every function has an inverse, which is in general a set-valued mapping that may not be a function. The classical inverse function theorem identifies the case when the inverse is actually a function, at least locally.

Recall that a linear and bounded mapping $A : X \to Y$ acting between Banach spaces is invertible when its inverse A^{-1} is single-valued, linear, and bounded as well. Also, recall that a generally set-valued mapping $F : X \rightrightarrows Y$ with $(\bar{x}, \bar{y}) \in \operatorname{gph} F$ is said to have a single-valued graphical localization s around $\bar{x}$ for $\bar{y}$ if there exist neighborhoods U of $\bar{x}$ and V of $\bar{y}$ such that $U \subset \operatorname{dom} s$ and $\operatorname{gph} s = (U \times V) \cap \operatorname{gph} F$.

Inverse Function Theorem. *Let X and Y be Banach spaces. Consider a function $f : X \to Y$ with $\bar{x} \in \operatorname{int} \operatorname{dom} f$ which is continuously Fréchet differentiable in a neighborhood of $\bar{x}$ and let the Fréchet derivative $Df(\bar{x})$ of f at $\bar{x}$ be invertible. Then the inverse f^{-1} at $\bar{y} := f(\bar{x})$ has a single-valued localization s around $\bar{y}$ for $\bar{x}$ which is continuously Fréchet differentiable in a neighborhood Q of $\bar{y}$ with derivative satisfying*

$$Ds(y) = Df(s(y))^{-1} \quad \textit{for every } y \in Q.$$

In particular, $Ds(\bar{y}) = Df(\bar{x})^{-1}$.

The classical implicit function theorem is concerned with the seemingly more general case, where an implicit function is defined by an equation $f(p, x) = 0$ where p is a parameter.

According to Stanimir Troyanski, this is the most important theorem in analysis.

© The Author(s), under exclusive license to Springer Nature Switzerland AG 2021
A. L. Dontchev, *Lectures on Variational Analysis*, Applied Mathematical Sciences 205, https://doi.org/10.1007/978-3-030-79911-3_8

Implicit Function Theorem. *Let X, Y, and P be Banach spaces. Consider a function $f : P \times X \to Y$ with $(\bar{p}, \bar{x}) \in \operatorname{dom} f$ such that $f(\bar{p}, \bar{x}) = 0$, which is continuously Fréchet differentiable in a neighborhood of $(\bar{p}, \bar{x})$, and suppose that the partial derivative mapping $D_x f(\bar{p}, \bar{x})$ is invertible. Then the solution mapping $p \mapsto S(p) = \{x \mid f(p, x) = 0\}$ has a single-valued localization s which is continuously differentiable around $\bar{p}$ and its derivative satisfies*

$$Ds(p) = -D_x f(p, s(p))^{-1} D_p f(p, s(p)) \quad \textit{for every } p \textit{ near } \bar{p}. \tag{1}$$

Observe that the inverse function theorem is a particular case of the implicit function theorem in which $f(p, x)$ has the form $-p + f(x)$ (with some abuse of notation). Actually, these two theorems are equivalent. We will show this equivalence in further lines, for simplicity in finite dimensions.

In the context of the statements of the inverse and implicit function theorems above, let $P = I\!R^d$ and $X = Y = I\!R^n$. Let I be the $d \times d$ identity matrix, 0 be the $d \times n$ zero matrix, B be an $n \times d$ matrix, and A be an $n \times n$ nonsingular matrix. We show first that the square matrix

$$J = \begin{pmatrix} I & 0 \\ B & A \end{pmatrix},$$

is nonsingular. Indeed, if J were singular, there exists $y \neq 0$ such that $Jy = 0$. For $y = (p, x) \in I\!R^d \times I\!R^n$ the equation $Jy = 0$ leads to $p = 0$ and $Ax = 0$. Hence there exists $x \neq 0$ with $Ax = 0$, which contradicts the nonsingularity of A. Now, consider the function

$$\varphi(p, x) = \begin{pmatrix} p \\ f(p, x) \end{pmatrix}$$

acting from $I\!R^d \times I\!R^n$ to itself. The inverse of this function is defined by the solutions of the equation

$$\varphi(p, x) = \begin{pmatrix} p \\ f(p, x) \end{pmatrix} = \begin{pmatrix} y_1 \\ y_2 \end{pmatrix},$$

where the vector $(y_1, y_2) \in I\!R^d \times I\!R^n$ is now the parameter and (p, x) is the dependent variable. The nonsingularity of the partial Jacobian $\nabla_x f(\bar{p}, \bar{x})$ implies through the auxiliary result above that the Jacobian of the function φ at the point $(\bar{x}, \bar{p})$,

$$J(\bar{p}, \bar{x}) = \begin{pmatrix} I & 0 \\ \nabla_p f(\bar{p}, \bar{x}) & \nabla_x f(\bar{p}, \bar{x}) \end{pmatrix},$$

is nonsingular as well. Then, according to the classical inverse function theorem, the inverse φ^{-1} has a single-valued localization

$$(y_1, y_2) \mapsto (q(y_1, y_2), r(y_1, y_2)) \text{ around } (\bar{p}, 0) \text{ for } (\bar{p}, \bar{x})$$

which is continuously differentiable around $(\bar{p}, 0)$. Note that

$$\begin{cases} q(y_1, y_2) & = y_1, \\ f(y_1, r(y_1, y_2)) = y_2. \end{cases}$$

Differentiating the second equality with respect to y_1 by using the chain rule, we get

$$\nabla_p f(y_1, r(y_1, y_2)) + \nabla_x f(y_1, r(y_1, y_2)) \cdot \nabla_{y_1} r(y_1, y_2) = 0.$$

When (y_1, y_2) is close to $(\bar{p}, 0)$, the point $(y_1, r(y_1, y_2))$ is close to $(\bar{p}, \bar{x})$ and then $\nabla_x f(y_1, r(y_1, y_2))$ is nonsingular. Thus, the last equation with respect to $\nabla_{y_1} r(y_1, y_2)$ gives

$$\nabla_{y_1} r(y_1, y_2) = -\nabla_x f(y_1, r(y_1, y_2))^{-1} \nabla_p f(y_1, r(y_1, y_2)).$$

In particular, at points $(y_1, y_2) = (p, 0)$ close to $(\bar{p}, 0)$ we conclude that the mapping $p \mapsto s(p) := r(p, 0)$ is a single-valued localization of the solution mapping S around $\bar{p}$ for $\bar{x}$ which is continuously differentiable around $\bar{p}$ and its derivative is given by (1).

The following exercise complements the classical implicit function theorem.

Exercise 8.1. *Let $f : I\!R^d \times I\!R^n \to I\!R^n$ be continuously differentiable in a neighborhood of $(\bar{p}, \bar{x})$ and such that $f(\bar{p}, \bar{x}) = 0$, and let $\nabla_p f(\bar{p}, \bar{x})$ be of full rank n. Suppose that the solution mapping $p \mapsto S(p) = \{x \mid f(p, x) = 0\}$ has a single-valued localization s around $\bar{p}$ for $\bar{x}$ which is continuously differentiable in a neighborhood Q of $\bar{p}$. Prove that the partial Jacobian $\nabla_x f(\bar{p}, \bar{x})$ must be nonsingular.*

Guide. Without loss of generality let $\bar{p} = 0$ and $\bar{x} = 0$ and let $A = \nabla_x f(0, 0)$ and $B = \nabla_p f(0, 0)$. Consider the function

$$\psi(q, x, y) = f(B^\mathsf{T} q, x) - Ax + y.$$

Observe that $\nabla_q \psi(0, 0, 0) = BB^\mathsf{T}$ is nonsingular since B is assumed surjective. Hence, by the classical implicit function theorem the solution mapping

$$I\!R^n \times I\!R^n \ni (x, y) \mapsto \Psi(x, y) := \left\{ q \in I\!R^n \,\middle|\, \psi(q, x, y) = 0 \right\}$$

has a single-valued smooth localization at $(0, 0)$ for 0; denote this localization by σ. Now assume that the solution mapping S has a single-valued smooth localization s at 0 for 0, i.e. $f(p, s(p)) = 0$ for all p near 0, s is smooth near 0 and $s(0) = 0$. Clearly, we could choose x and y so close to 0 that the norms of $p = B^\mathsf{T} q$, $q = \sigma(x, y)$ and $s(p)$ are small enough to satisfy $q = \sigma(x, y) \in \Psi(x, y)$ and $x = s(B^\mathsf{T} q) \in S(B^\mathsf{T} q)$. Then $s(B^\mathsf{T} \sigma(x, y))$ satisfies $-As(B^\mathsf{T} \sigma(x, y)) + y = 0$. We obtain that for any $y \in I\!R^n$ (not necessarily close to zero, by linearity) the equation $Ax = y$ has a solution. Since A is a square matrix, it must be nonsingular. □

We will now introduce a new notion of regularity of mappings, which plays a major role in optimization and beyond. It is an analog of invertibility now adapted for set-valued mappings and metric spaces.

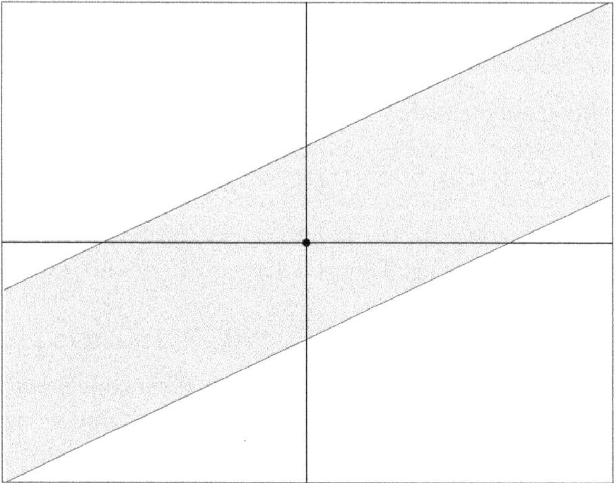

Fig. 8.1: Graph of a mapping which is metrically regular at each point of its graph, by the Robinson–Ursescu theorem, but not strongly regular anywhere

Strong Regularity. *A mapping $F : X \to Y$ acting between metric spaces X and Y is said to be* strongly regular *at $\bar{x}$ for $\bar{y}$ when $(\bar{x}, \bar{y}) \in \operatorname{gph} F$ and the inverse F^{-1} has a Lipschitz continuous single-valued localization around $\bar{y}$ for $\bar{x}$.*

For a linear mapping in finite dimensions represented by a matrix A, strong regularity coincides with the nonsingularity of A and requires A be a square matrix. For any function $f : I\!R^n \to I\!R^m$, strong regularity requires $m = n$, by the Brouwer invariance of domain theorem, which we state next.

Invariance of Domain Theorem (Brouwer). *Let $O \subset I\!R^n$ be open and for $m \le n$ let $f : O \to I\!R^m$ be continuous and such that f^{-1} is single-valued on $f(O)$. Then $f(O)$ is open, f^{-1} is continuous on $f(O)$, and $m = n$.*

Strong regularity is also called *strong metric regularity* which points to a connection with metric regularity (cf, Fig. 8.1). The following proposition relates the existence of a Lipschitz continuous single-valued localization with the Aubin property.

Proposition 8.2. *For a set-valued mapping $S : X \rightrightarrows Y$ acting between metric spaces and a pair $(\bar{y}, \bar{x}) \in \operatorname{gph} S$, the following are equivalent:*

(a) *S has a Lipschitz continuous single-valued localization s around $\bar{y}$ for $\bar{x}$.*

(b) *S has the Aubin property at $\bar{y}$ for $\bar{x}$ as well as a localization around $\bar{y}$ for $\bar{x}$ that is nowhere multivalued.*

Moreover, under the conditions in (b) the localization s in (a) has

$$\operatorname{lip}(s; \bar{y}) = \operatorname{lip}(S; \bar{y} \,|\, \bar{x}).$$

Proof. The existence of a Lipschitz single-valued localization in (a) implies that S has the Aubin property at $\bar{y}$ for $\bar{x}$ with $\operatorname{lip}(S;\bar{y}\,|\,\bar{x}) \le \operatorname{lip}(s;\bar{y})$. Then, by Proposition 4.1, the domain of S contains a neighborhood of $\bar{y}$.

In the opposite direction, condition (b) is equivalent to the property that S has a localization σ around $\bar{y}$ for $\bar{x}$ that is not multivalued and is Lipschitz continuous around $\bar{y}$ with Lipschitz constant any $\kappa > \operatorname{lip}(S;\bar{y}\,|\,\bar{x})$. Let a and b be positive constants such that $y \mapsto \sigma(y) := S(y) \cap I\!B_a(\bar{x})$ is defined on $I\!B_b(\bar{y})$ and let $b' > 0$ satisfy $b' < \min\{b, a/8\kappa\}$. Then for $y, y' \in I\!B_{b'}(\bar{y})$ we have

$$\begin{aligned} d(\sigma(y), S(y')) &= d(S(y) \cap I\!B_a(\bar{x}), S(y')) \\ &\le \kappa\rho(y, y') \le \kappa\rho(y, \bar{y}) + \kappa\rho(y', \bar{y}) \le 2\kappa b' < a/4. \end{aligned}$$

Hence, there exists $x' \in S(y')$ such that $\rho(x', s(y)) \le d(\sigma(y), S(y')) + a/4 < a/2$. Since $\rho(\sigma(y), \bar{x}) = d(\bar{x}, S(y))$ and

$$\rho(x', \bar{x}) \le \rho(x', \sigma(y)) + \rho(s(y), \bar{x}) < a/2 + d(\bar{x}, S(y)) \le a/2 + \kappa\rho(y, \bar{y}) < a,$$

we obtain that

$$d(\sigma(y), S(y') \cap I\!B_a(\bar{x})) = d(\sigma(y), S(y')).$$

Then

$$\begin{aligned} \kappa\rho(y, y') &\ge d(S(y) \cap I\!B_a(\bar{x}), S(y')) \\ &= d(\sigma(y), S(y')) = d(\sigma(y), S(y') \cap I\!B_a(\bar{x})) = \rho(\sigma(y), \sigma(y')). \end{aligned}$$

Thus, σ is a Lipschitz continuous single-valued localization of S around $\bar{y}$ for $\bar{x}$ with Lipschitz constant κ, hence $\operatorname{lip}(\sigma;\bar{y}) \le \operatorname{lip}(S;\bar{y}\,|\,\bar{x})$. This completes the proof. □

In a path-breaking work, S. M. Robinson [67] extended the classical implicit function theorem to variational inequalities. We will present here the following version of this result:

Theorem 8.3 (Robinson). *Let P and X be Banach spaces and let X^* be the dual for X. For a function $f : P \times X \to X^*$ and a closed and convex set C in X, consider the parameterized variational inequality to find $x \in C$ such that*

$$\langle f(p, x), y - x\rangle \ge 0 \quad \textit{for all} \quad y \in C,$$

or equivalently,

$$f(p, x) + N_C(x) \ni 0,$$

where $p \in P$ is a parameter, and $N_C : X \to X^*$ is defined as the usual normal cone mapping. Denote by S its solution mapping and let $\bar{x} \in S(\bar{p})$. Assume that:

(a) *$f(p, x)$ is Fréchet differentiable with respect to x in a neighborhood of the point $(\bar{p}, \bar{x})$, and both $f(p, x)$ and $D_x f(p, x)$ depend continuously on (p, x) in this neighborhood. Also, f is Lipschitz continuous in p uniformly in x around $(\bar{p}, \bar{x})$;*

(b) *The inverse G^{-1} of the set-valued mapping $G : X \rightrightarrows X^*$ defined as*

$$x \mapsto G(x) = f(\bar{p}, \bar{x}) + D_x f(\bar{p}, \bar{x})(x - \bar{x}) + N_C(x), \qquad \text{with } G(\bar{x}) \ni 0,$$

has a Lipschitz continuous single-valued localization around 0 *for* $\bar{x}$.

Then S *has a single-valued localization* s *around* $\bar{p}$ *for* $\bar{x}$ *which is Lipschitz continuous around* $\bar{p}$.

Note that, in relation to the classical implicit function theorem, Robinson theorem considers the solution mapping of a variational inequality, not just an equation. In the case when $C = X$, this variational inequality becomes an equation, and moreover, the condition (b) for the mapping G reduces to nonsingularity of $D_x f(\bar{p}, \bar{x})$, as in the classical case. On the other hand, the solution mapping is claimed to have a single-valued localization which is merely Lipschitz continuous but may be not differentiable, at least in the usual sense. Indeed, consider the variational inequality as above with $P = X = \mathbb{R}$, $f(p, x) = p - x$, and $C = [0, +\infty)$. Its solution is $x(p) = 0$ for $p \le 0$ and $x(p) = p$ for $p > 0$, a Lipschitz continuous function which is not differentiable at $p = 0$.

Exercise 8.4. *Derive the classical implicit function theorem from Theorem 8.3.*

Hint. In order to show continuous differentiability of the Lipschitz localization s whose existence follows from Theorem 8.3, use the equality

$$\begin{aligned} s(p+h) - s(p) - D_x f(p, s(p))^{-1} h \\ = -D_x f(p, s(p))^{-1}(f(s(p+h)) - f(p, s(p)) \\ -D_x f(p, s(p))(s(p+h) - s(p))). \end{aligned}$$

□

The following theorem is an extended version of (Robinson) Theorem 8.3. Note that it is stated for generalized equations in metric spaces.

Theorem 8.5. *Let* X *be a complete metric space, let* Y *be a linear metric space with a shift-invariant metric, and let* P *be a metric space. For a function* $f : P \times X \to Y$ *and a set-valued mapping* $F : X \rightrightarrows Y$*, consider the generalized equation*

$$f(p, x) + F(x) \ni 0 \tag{2}$$

with solution mapping

$$S(p) = \left\{ x \,\middle|\, f(p, x) + F(x) \ni 0 \right\} \quad \text{having } \bar{x} \in S(\bar{p}). \tag{3}$$

Let κ *and* μ *be positive constants such that* $\kappa\mu < 1$*. Let* $h : X \to Y$ *be a function which satisfies the conditions*

$$h(\bar{x}) = f(\bar{p}, \bar{x}) \quad \text{and} \quad \widehat{\text{lip}}_x(f - h; (\bar{p}, \bar{x})) \le \mu, \tag{4}$$

and suppose that $h + F$ *is strongly regular at* $\bar{x}$ *for* 0 *or, equivalently, the inverse* $(h + F)^{-1}$ *has a Lipschitz continuous single-valued localization* ω *around* 0 *for* $\bar{x}$*; moreover, let*

$$\operatorname{reg}(h + F; \bar{x} \,|\, 0) = \operatorname{lip}(\omega; 0) \le \kappa.$$

Then the mapping S has a single-valued localization s around $\bar{p}$ for $\bar{x}$ which is Lipschitz continuous near $\bar{p}$ with

$$\operatorname{lip}(s; \bar{p}) \le \frac{\kappa}{1 - \kappa\mu} \widehat{\operatorname{lip}}_p(f; (\bar{p}, \bar{x})). \tag{5}$$

Proof. According to Proposition 8.2, the mapping $h + F$ is metrically regular at $\bar{x}$ for 0, hence, from Theorem 4.3 the mapping S has the Aubin property at $\bar{p}$ for $\bar{x}$. Hence, it is sufficient to show that there is a localization of S around $\bar{y}$ for $\bar{x}$ that is nowhere multivalued. We will show that the assumption that every localization of S around $\bar{p}$ for $\bar{x}$ is multivalued somewhere around $\bar{x}$ leads to a contradiction.

First, observe that

$$x \in S(p) \iff x \in (f(p, \cdot) + F(\cdot))^{-1}(0) = (h + F)^{-1}(h(x) - f(p, x)).$$

The assumptions of the theorem yield that there exist $a > 0$ and $b > 0$ such that the function $y \mapsto \omega(y) := (h + F)^{-1}(y) \cap \mathbb{B}_a(\bar{x})$ is Lipschitz continuous on $\mathbb{B}_b(0)$ with constant κ. Choose $c > 0$ such that

$$\rho(h(x) - f(p, x), h(x') - f(p, x')) \le \mu\rho(x, x') \quad \text{for all} \quad x, x' \in \mathbb{B}_a(\bar{x}) \text{ and } p \in \mathbb{B}_c(\bar{p}).$$

Make a smaller if necessary so that $2\mu a \le b$.

Suppose that there exists $p \in \mathbb{B}_c(\bar{p})$ and $x, x' \in \mathbb{B}_a(\bar{x})$, $x \ne x'$, such that

$$\rho(f(p, x) - h(x), 0) \le \mu\rho(x, x') \le \mu(2a) \le b.$$

Then we have

$$x = (h + F)^{-1}(h(x) - f(p, x)) \quad \text{and} \quad x' = (h + F)^{-1}(h(x') - f(p, x')),$$

therefore

$$\begin{aligned} \rho(x, x') &= \rho((h + F)^{-1}(h(x) - f(p, x)), (h + F)^{-1}(h(x') - f(p, x')) \\ &\le \kappa\rho(h(x) - f(p, x), h(x') - f(p, x')) \le \kappa\mu\, \rho(x, x') < \rho(x, x'), \end{aligned}$$

which is impossible. Hence there exists a localization of S around $\bar{y}$ for $\bar{x}$ that is nowhere multivalued. From Proposition 8.2 we obtain that the mapping S has a Lipschitz continuous single-valued localization s around $\bar{p}$ for $\bar{x}$. Clearly,

$$s(p) = \omega(e(p, s(p))) \quad \text{where} \quad e(p, x) = h(x) - f(p, x).$$

It remains to show (5).

Assume that $\widehat{\operatorname{lip}}_p(f; (\bar{p}, \bar{x})) < \gamma$ for some $\gamma > 0$ and choose any $p', p \in \mathbb{B}_c(\bar{p})$. We have

$$\begin{aligned}\rho(s(p'), s(p)) &= \rho(\omega(e(p', s(p'))), \omega(e(p, s(p)))) \\ &\le \rho(\omega(e(p', s(p'))), \omega(e(p', s(p)))) \\ &\qquad + \rho(\omega(e(p', s(p))), \omega(e(p, s(p)))) \\ &\le \kappa\, \rho(e(p', s(p')), e(p', s(p))) \\ &\qquad + \kappa\, \rho(e(p', s(p)), e(p, s(p))) \\ &\le \kappa\mu\, \rho(s(p'), s(p)) + \kappa\gamma\rho(p, p').\end{aligned}$$

Rearranging the last inequality and passing to the lower bound with γ yields (5) and completes the proof. □

Note the similarity of Theorem 8.5 with Theorem 5.5. In particular, Theorem 8.5 shows that strong regularity obeys the same paradigm as metric regularity: *if a mapping F is strongly regular, then for any function g with a sufficiently small Lipschitz modulus the mapping $f + F$ is strongly regular as well.* In the case when $f(p, x) = -p + g(x)$, where $P = Y$ and the function $g : X \to Y$, from Theorem 8.5 we obtain a result which parallels Theorem 5.2:

Theorem 8.6. *Let X be a complete metric space and let Y be a linear metric space with shift-invariant metric. Consider a set-valued mapping $F : X \rightrightarrows Y$, a point $(\bar{x}, \bar{y}) \in \operatorname{gph} F$ and a function $g : X \to Y$ with $\bar{x} \in \operatorname{int} \operatorname{dom} g$. Let F be strongly regular at $\bar{x}$ for $\bar{y}$ with a Lipschitz constant κ of the single-valued localization of F^{-1} around $\bar{y}$. Also, let μ be nonnegative constant such that*

$$\kappa\mu < 1, \qquad \text{and} \qquad \operatorname{lip}(g; \bar{x}) \le \mu.$$

Then the mapping $g + F$ is strongly regular at $\bar{x}$ for $g(\bar{x}) + \bar{y}$ with a Lipschitz modulus of the single-valued localization s of $(g + F)^{-1}$ satisfying

$$\operatorname{lip}(s; g(\bar{x}) + \bar{y}) \le \frac{\kappa}{1 - \kappa\mu}.$$

Analogously, the following corollary parallels Corollary 5.3:

Corollary 8.7. *Let X and Y be Banach spaces. For a function $g : X \to Y$ which is strictly differentiable at $\bar{x}$ and a set-valued mapping $F : X \rightrightarrows Y$, the following are equivalent:*

(i) The mapping $g + F$ is strongly regular at $\bar{x}$ for $\bar{y}$ with a single-valued localization s of the inverse;

(ii) The mapping $x \mapsto g(\bar{x}) + Dg(\bar{x})(x - \bar{x}) + F(x)$ is strongly regular at $\bar{x}$ for $\bar{y}$ with a single-valued localization σ of the inverse.

Moreover,

$$\operatorname{lip}(s; \bar{y}) = \operatorname{lip}(\sigma; \bar{y}).$$

We present next an important addition to Theorem 8.5 which shows that not only the existence of a single-valued localization of the inverse which is Lipschitz continuous is inherited by a mapping from its approximation represented by, e.g., linearization, but also, under slightly stronger conditions, a stronger property of differentiability of this localization is inherited as well. Specifically, we have

Theorem 8.8. *In the setting of Theorem 8.5, assume that X, Y, and P are Banach spaces and the constant μ is zero. Then, with the localization ω of $(h + F)^{-1}$, we have the following additions to the conclusions of Theorem 8.5:*

If $\widehat{\text{lip}}_p(f;(\bar{p},\bar{x})) < \infty$, then the single-valued localization s of S is Lipschitz continuous near $\bar{p}$ with

$$\text{lip}\,(s;\bar{p}) \le \text{lip}\,(\omega;0)\cdot\widehat{\text{lip}}_p(f;(\bar{p},\bar{x})). \tag{6}$$

In addition, if there is a function $r : P \to Y$ such that

$$\text{clm}_p(f - r;(\bar{p},\bar{x})) = 0,$$

then the function $\eta : P \to X$ defined by

$$\eta(p) = \omega(-r(p) + f(\bar{p},\bar{x})) \tag{7}$$

satisfies

$$\text{clm}\,(\eta - s;\bar{p}) = 0. \tag{8}$$

Proof. Let the constants a and c be as in the proof of Theorem 8.5, $Q = I\!B_c(\bar{p})$ and $U = I\!B_a(\bar{x})$. For $p \in Q$ and $x \in U$, we have

$$x \in (h + F)^{-1}(h(x) - f(p,x)) \iff x \in S(p).$$

Hence, for $e(p,x) = f(p,x) - h(x)$,

$$s(p) = \omega(-e(p,s(p))) \;\text{ and }\; \bar{x} = s(\bar{p}) = \omega(0). \tag{9}$$

Then, taking into account that $\mu = 0$, (6) follows from (5) with $\kappa = \text{lip}\,(\omega;0)$.

Consider now any $\lambda > \text{clm}\,(s;\bar{p})$, any $\mu > \text{lip}\,(\omega;\bar{x})$ and any $\varepsilon > 0$. Make the neighborhoods Q and U smaller if necessary so that for all $p \in Q$ and $x \in U$ we have

$$\|s(p) - s(\bar{p})\| \le \lambda\|p - \bar{p}\|, \quad \|e(p,x) - e(p,\bar{x})\| \le \varepsilon\|x - \bar{x}\| \tag{10}$$

and

$$\|f(p,\bar{x}) - r(p)\| \le \varepsilon\|p - \bar{p}\|, \tag{11}$$

and furthermore so that the points $-e(p,x)$ and $-r(p) + f(\bar{p},\bar{x})$ are contained in a neighborhood of 0 on which the function ω is Lipschitz continuous with Lipschitz constant μ. Then, for $p \in Q$, using (7), (9), (10), and (11) and the fact that $e(\bar{p},\bar{x}) = 0$, we get

$$\begin{aligned}\|s(p) - \eta(p)\| &= \|s(p) - \omega(-r(p) + f(\bar{p}, \bar{x}))\| \\ &= \|\omega(-e(p, s(p))) - \omega(-r(p) + f(\bar{p}, \bar{x}))\| \\ &\leq \mu(\| - e(p, s(p)) + e(p, \bar{x})\| + \|f(p, \bar{x}) - r(p)\|) \\ &\leq \mu\varepsilon\|s(p) - \bar{x}\| + \mu\varepsilon\|p - \bar{p}\| \leq \varepsilon\mu(\lambda + 1)\|p - \bar{p}\|.\end{aligned}$$

Since ε can be arbitrarily small and also $s(\bar{p}) = \bar{x} = \omega(0) = \eta(\bar{p})$, the function η defined in (7) satisfies (8). □

Recall that a function $s : P \to X$ is semidifferentiable at $\bar{p}$ if there exists a continuous and positively homogeneous function $Ds(\bar{p}) : P \to X$ such that

$$\operatorname{clm}(e; \bar{p})) = 0 \quad \text{for} \quad e(p) = s(p) + Ds(\bar{p})(p - \bar{p}).$$

From Theorem 8.8, we obtain the following corollary which extends Corollary 8.7:

Corollary 8.9. *Consider the generalized equation* (2) *with solution mapping S given by* (3) *and suppose that f is strictly differentiable at $(\bar{p}, \bar{x})$. Assume that the inverse G^{-1} of the mapping*

$$G(x) = f(\bar{p}, \bar{x}) + \nabla_x f(\bar{p}, \bar{x})(x - \bar{x}) + F(x), \qquad \text{with } G(\bar{x}) \ni 0,$$

has a Lipschitz continuous single-valued localization ω around 0 for $\bar{x}$ which is positively homogeneous. Then not only do the conclusions of Theorem 8.8 hold for a solution localization s with (6) *becoming*

$$\operatorname{lip}(s; \bar{p}) \leq \operatorname{lip}(\omega; 0) \cdot \|\nabla_p f(\bar{p}, \bar{x})\|,$$

but also s is semidifferentiable at $\bar{p}$ with semiderivative given by the function

$$Ds(\bar{p})(q) = \omega\big(- \nabla_p f(\bar{p}, \bar{x})(q)\big). \tag{12}$$

Proof. In this case, Theorem 8.8 is applicable with h taken to be the linearization of $f(\bar{p}, \cdot)$ at $\bar{x}$ and r taken to be the linearization of $f(\cdot, \bar{x})$ at $\bar{p}$. It remains to use (8) and the positive homogeneity of ω to show that the function defined in (12) is indeed the semiderivative of s at $\bar{p}$. □

In the following lecture we will show that, for a variational inequality over a polyhedral set, the approximating localization of the inverse of the mapping G is positively homogeneous and then from Corollary 8.9 we obtain semidifferentiability of a localization of the solution mapping.

Lecture 9
Variational Inequalities over Polyhedral Sets

In Lecture 1 we defined polyhedral sets as intersections of finitely many half-spaces and also studied there the geometry of various cones to such sets, such as the tangent, normal, polar, and critical cone. Let C be a polyhedral set in $I\!R^n$ represented by

$$C = \{x \mid \langle b_i, x\rangle \leq \alpha_i \text{ for } i = 1, \ldots m\} \tag{1}$$

for some $b_i \in I\!R^n$ and $\alpha_i \in I\!R$, $i = 1, \ldots, m$. For an $x \in C$ the set of indices of the constraints in (1) that are active at x is $I(x) = \{ i \,|\, \langle b_i, x\rangle = \alpha_i \}$. Polyhedral cones are characterized by having a representation (1) in which $\alpha_i = 0$ for all i. The polyhedral cones is that they can equally well be represented as a collection of vectors $b_1, \ldots, b_m$ such that

$$K = \{ y_1 b_1 + \cdots + y_m b_m \,|\, y_i \geq 0 \text{ for } i = 1, \ldots, m \}. \tag{2}$$

This fact is known in the literature as the Minkowski–Weyl theorem. The following proposition gives a description of the polar to a convex closed cone which is not necessarily polyhedral.

Proposition 9.1. *Let K be a closed, convex cone in $I\!R^n$ and let K^* be its polar defined by*

$$K^* = \{ y \,|\, \langle x, y\rangle \leq 0 \textit{ for all } x \in K \}. \tag{3}$$

Then K^* is a closed, convex cone, and its polar $(K^*)^*$ is in turn K. Furthermore, the normal vectors to K and K^* are related by

$$y \in N_K(x) \iff x \in N_{K^*}(y) \iff x \in K,\ y \in K^*,\ \langle x, y\rangle = 0. \tag{4}$$

Proof. First consider any $x \in K$ and $y \in N_K(x)$. From the definition of polarity in (3) we have $\langle y, x' - x\rangle \leq 0$ for all $x' \in K$, so the maximum of $\langle y, x'\rangle$ over $x' \in K$ is attained at x. Because K contains all positive multiples of each of its vectors, this comes down to having $\langle y, x\rangle = 0$ and $\langle y, x'\rangle \leq 0$ for all $x' \in K$. Therefore $N_K(x) = \{ y \in K^* \,|\, y \perp x \}$.

© The Author(s), under exclusive license to Springer Nature Switzerland AG 2021

A. L. Dontchev, *Lectures on Variational Analysis*, Applied Mathematical Sciences 205, https://doi.org/10.1007/978-3-030-79911-3_9

It is elementary that K^* is a cone which is closed and convex, with $(K^*)^* \supset K$. Consider any $z \notin K$. Let $x = P_K(z)$ and $y = z - x$. Then $y \neq 0$ and $P_K(x + y) = x$, hence $y \in N_K(x)$, so that $y \in K^*$ and $y \perp x$. We have $\langle y, z\rangle = \langle y, y\rangle > 0$, which confirms that $z \notin (K^*)^*$. Therefore $(K^*)^* = K$. The formula for normals to K must hold then equally for K^* through symmetry: $N_{K^*}(y) = \{ x \in K \,|\, x \perp y \}$ for any $y \in K^*$. This establishes the normality relations (4). □

Proposition 9.2. *Let C be a polyhedral set represented as in* (1) *and let $x \in C$. Then the tangent and normal cones to C at x are polyhedral cones, with the tangent cone having the representation*

$$T_C(x) = \{ w \,|\, \langle b_i, w\rangle \le 0 \text{ for } i \in I(x) \} \tag{5}$$

and the normal cone having the representation

$$N_C(x) = \left\{ v \,\middle|\, v = \sum\nolimits_{i=1}^{m} y_i b_i \text{ with } y_i \ge 0 \text{ for } i \in I(x),\ y_i = 0 \text{ for } i \notin I(x) \right\}. \tag{6}$$

Furthermore, the tangent cone has the properties that

$$W \cap [C - x] = W \cap T_C(x) \text{ for some neighborhood } W \text{ of } 0 \tag{7}$$

and

$$T_C(x) \supset T_C(\bar{x}) \text{ for all } x \text{ in some neighborhood } U \text{ of } \bar{x}. \tag{8}$$

Proof. The formula (5) for $T_C(x)$ follows from (1) just by applying the definition of the tangent cone. Then from (5) and the preceding facts about polyhedral cones and polarity, utilizing also (4), we obtain (6). The equality (7) is deduced simply by comparing (1) and (5). To obtain (8), observe that $I(x) \subset I(\bar{x})$ for x close to $\bar{x}$ and then the inclusion follows from (5). □

Recall that for a convex set C, any $x \in C$ and any $v \in N_C(x)$, the critical cone to C at x for v is given by

$$K_C(x, v) = \{ w \in T_C(x) \,|\, w \perp v \}.$$

If C is polyhedral, then $K_C(x, v)$ is polyhedral as well, as seen immediately from the representation in (5). An important result about these cones is the so-called Reduction lemma we give next.

Reduction Lemma. *Let C be a polyhedral set in $I\!R^n$, and let $\bar{x} \in C$, $\bar{v} \in N_C(\bar{x})$ and $K = K_C(\bar{x}, \bar{v})$ be the corresponding critical cone. Then there is a neighborhood O of the origin* $(0, 0)$ *such that*

$$O \cap [\operatorname{gph} N_C - (\bar{x}, \bar{v})] = O \cap \operatorname{gph} N_K.$$

In other words, one has

$$\bar{v} + u \in N_C(\bar{x} + w) \iff u \in N_K(w) \text{ for } (w, u) \text{ near } (0, 0). \tag{9}$$

We employ next Reduction Lemma to characterize strong regularity of an affine variational inequality.

Theorem 9.3. *Consider the variational inequality*

$$a + Ax + N_C(x) \ni 0,$$

where $a \in I\!R^n$, A is an $n \times n$ matrix, and C is a polyhedral set in $I\!R^n$. Let $\bar{x}$ be a solution and let $\bar{v} = a - A\bar{x}$, so that $\bar{v} \in N_C(\bar{x})$, and let $K = K_C(\bar{x}, \bar{v})$ be the associated critical cone. Then the mapping

$$x \mapsto G(x) = a + Ax + N_C(x) \text{ with } G(\bar{x}) \ni 0$$

is strongly regular at $\bar{x}$ for 0 if and only if the inverse G_0^{-1} of the mapping

$$w \mapsto G_0(w) = Aw + N_K(w) \text{ with } G_0(0) \ni 0$$

is a single-valued Lipschitz continuous mapping with all of $I\!R^n$ as its domain, and the function $\sigma(v) = \bar{x} + G_0^{-1}(v)$ is a localization of G^{-1} at 0 for $\bar{x}$.

Proof. According to Reduction Lemma, for (w, u) in some neighborhood of $(0, 0)$ we have that $\bar{v} + u \in N_C(\bar{x} + w)$ if and only if $u \in N_K(w)$. In the change of notation from u to $v = u + Aw$, this means that, for (w, v) in a neighborhood of $(0, 0)$, we have $v \in G(\bar{x} + w)$ if and only if $v \in G_0(w)$. Thus, strong regularity of G at $\bar{x}$ for 0, which is equivalent to the existence of a Lipschitz continuous single-valued localization σ of G^{-1} around 0 for $\bar{x}$, corresponds to the existence of a Lipschitz continuous single-valued localization σ_0 of G_0^{-1} around 0 for 0; the relationship is given by $\sigma(v) = \bar{x} + \sigma_0(v)$. But whenever $v \in G_0(w)$ we have $\lambda v \in G_0(\lambda w)$ for all $\lambda > 0$, i.e., the graph of G_0 is a cone. Therefore, when σ_0 exists it can be scaled arbitrarily large and must correspond to G_0^{-1} being a Lipschitz continuous single-valued mapping with all of $I\!R^n$ as its domain. □

In fact, if G_0^{-1} is single-valued globally, it is automatically globally Lipschitz continuous. This is to be shown in the following exercise.

Exercise 9.4. *Prove that, for an $n\times n$ matrix A and a polyhedral set C, if the mapping $(A + N_C)^{-1}$ is single-valued, then it is Lipschitz continuous.*

Hint. First note that the mapping $A+N_C$ is piecewise polyhedral and so is its inverse, see Exercise 3.5. Then apply Theorem 3.4 to obtain that the inverse is outer Lipschitz continuous. To finish the proof, it is enough to show that the inverse $(A + N_C)^{-1}$ is inner semicontinuous, hence continuous. □

Consider now the variational inequality

$$f(p, x) + N_C(x) \ni 0, \tag{10}$$

where the function $f : I\!R^d \times I\!R^n \to I\!R^n$ is strictly differentiable at $(\bar{p}, \bar{x})$ and the set C is polyhedral. Theorem 8.3 tells us that the strong regularity of the linearization

$$G(x) = f(\bar{p}, \bar{x}) + \nabla_x f(\bar{p}, \bar{x})(x - \bar{x}) + N_C(x), \quad \text{with } G(\bar{x}) \ni 0$$

of the mapping on the left side of (10) is equivalent to the critical cone $K = K_C(\bar{x}, -f(\bar{p}, \bar{x}))$ being such that the inverse G_0^{-1} of the mapping $G_0(w) = \nabla_x f(\bar{x}, \bar{p})w + N_K(w)$ is single-valued and Lipschitz continuous from all of $I\!R^n$ into itself. In putting these facts together with observations about the special nature of G_0, we obtain the following powerful fact about variational inequalities over polyhedral sets.

Theorem 9.5. *For the variational inequality* (10) *under the assumption that C is polyhedral and f is strictly differentiable at $(\bar{p}, \bar{x})$, with $\bar{x} \in S(\bar{p})$, let*

$$A = \nabla_x f(\bar{p}, \bar{x}) \quad \text{and} \quad K = K_C(\bar{x}, \bar{v}) \text{ for } \bar{v} = -f(\bar{p}, \bar{x}).$$

Suppose that for each $u \in I\!R^n$ there is a unique solution $w(u)$ to the auxiliary variational inequality $Aw - u + N_K(w) \ni 0$, this being equivalent to saying that the mapping

$$\omega := (A + N_K)^{-1} \text{ is everywhere single-valued,}$$

in which case ω is Lipschitz continuous globally. Then the solution mapping of (10) *has a Lipschitz continuous single-valued localization s around $\bar{p}$ for $\bar{x}$ which satisfies*

$$\operatorname{lip}(s; \bar{p}) \leq \operatorname{lip}(\omega; 0) \cdot \|\nabla_p f(\bar{p}, \bar{x})\|.$$

Furthermore, s is semidifferentiable at $\bar{p}$ with semiderivative given by

$$Ds(\bar{p})(q) = \omega(-\nabla_p f(\bar{p}, \bar{x})q). \tag{11}$$

Proof. We merely have to combine the observation made before this theorem's statement with the statement of Corollary 8.9. Because K is a cone, the mapping N_K is positively homogeneous, and the same is true then for $A + N_K$ and its inverse. Thus, the function $Ds(\bar{p})$ in (11) furnishes the semiderivative at $\bar{p}$ of s. □

We apply Theorem 9.5 to the following optimization problem:

$$\text{minimize } g(p, x) \text{ over all } x \in C, \tag{12}$$

where $g : I\!R^d \times I\!R^n \to I\!R$ is twice continuously differentiable and C is a nonempty polyhedral set in $I\!R^n$. The variational inequality

$$\nabla_x g(p, x) + N_C(x) \ni 0 \tag{13}$$

provides for each p a first-order condition which x must satisfy if it furnishes a local minimum, but only describes, in general, the stationary points x for the minimization in (12). The basic object of interest to us for now, however, is the *stationary point mapping* $S : I\!R^d \rightrightarrows I\!R^n$ defined by

$$S(p) = \{ x \mid \nabla_x g(p, x) + N_C(x) \ni 0 \}.$$

The stationary point mapping for the linearization of (13) is accordingly the mapping $\bar{S} : I\!R^n \rightrightarrows I\!R^n$ defined by

$$\bar{S}(v) = \left\{ w \,\middle|\, \nabla_x g(\bar{p}, \bar{x}) + \nabla^2_{xx} g(\bar{p}, \bar{x})w + N_W(w) \ni v \right\},$$

where $W = C - \bar{x}$ and then $N_W(w) = N_C(\bar{x} + w)$. The following statement translates Theorem 9.5 to the setting of problem (12):

Theorem 9.6. *Suppose in the preceding notation, with $\bar{x} \in S(\bar{p})$, that the mapping $\bar{S}$ has a Lipschitz continuous single-valued localization $\bar{s}$ around 0 for 0. Then S has a Lipschitz continuous single-valued localization s around $\bar{p}$ for $\bar{x}$ with*

$$\text{lip}\,(s; \bar{p}) \leq \text{lip}\,(\bar{s}; 0) \cdot \|\nabla^2_{xp} g(\bar{p}, \bar{x})\|.$$

Furthermore, the localization s is semidifferentiable at $\bar{p}$ with semiderivative given by

$$Ds(\bar{p})(q) = \bar{s}(-\nabla^2_{xp} g(\bar{p}, \bar{x})q).$$

We already mentioned that strong regularity implies metric regularity. Indeed, Lipschitz continuity of a single-valued localization of the inverse of a mapping implies the Aubin property of the inverse, which is equivalent to metric regularity of the mapping. In some situations, metric regularity automatically entails strong regularity, that is, these two properties are equivalent. That is the case, for instance, for a linear mapping from $I\!R^n$ to itself represented by an $n \times n$ matrix A. Indeed, A is metrically regular if and only if it is surjective, which is the same as saying that A is nonsingular, so that we have strong regularity. More generally, we have the following:

Theorem 9.7. *For a polyhedral set C in $I\!R^n$ with a normal cone mapping N_C and an $n \times n$ matrix A, the mapping $A + N_C$ is strongly regular at $\bar{x}$ for $\bar{y}$ if and only of it is metrically regular there.*

We will not give a proof of this result here, see the bibliographical remarks for this lecture. An important corollary of this theorem is that for the mapping describing the Karush–Kuhn–Tucker optimality system in a nonlinear programming problem, metric regularity is the same as strong regularity.

We will describe next a class of mappings for which metric regularity implies strong regularity. This class depends on a localized, set-valued form of the monotonicity concept. To do that we consider mappings acting in Hilbert spaces.

Local Monotonicity. *Let H be a Hilbert space. A mapping $F : H \rightrightarrows H$ is said to be locally monotone at $\bar{x}$ for $\bar{y}$ if $(\bar{x}, \bar{y}) \in$ gph F and for some neighborhood W of $(\bar{x}, \bar{y})$, one has*

$$\langle y' - y, x' - x \rangle \geq 0 \quad \textit{whenever } (x', y'), (x, y) \in \text{gph } F \cap W.$$

In addition, if there exists a positive constant μ such that

$$\langle y' - y, x' - x \rangle \geq \mu \|x - x'\|^2 \quad \text{whenever } (x', y'), (x, y) \in \operatorname{gph} F \cap W,$$

then the mapping F is *strongly locally monotone*. When $W = H$ then the mapping F is just monotone or strongly monotone.

For an example of a monotone mapping, consider the variational inequality

$$\langle Ax + p, v - x \rangle \geq 0 \quad \text{for every } v \in C,$$

where $A : H \to H$ is a linear and bounded mapping which is self-adjoint, H is a Hilbert space as before, and C is a closed and convex set in H, and $p \in H$ is a parameter. We already know that the variational inequality can be written in terms of the normal cone mapping N_C as follows:

$$Ax + p + N_C(x) \ni 0.$$

Suppose that there exists a constant $\alpha > 0$ such that

$$\langle A(x - x'), x - x' \rangle \geq \alpha \|x - x'\|^2 \quad \text{for every } x, x' \in C,$$

which can be also written as

$$\langle Ax, x \rangle \geq \alpha \|x\|^2 \quad \text{for every } x \in C - C, \tag{14}$$

where $C - C = \{x \mid \exists u, v \in C, x = u - v\}$ is the usual algebraic difference of sets. We will show next that in this case the mapping $A + N_C$ is strongly monotone.

Let $x, x' \in C$ and let $v \in Ax + N_C(x)$, $v' \in Ax' + N_C(x')$. Then

$$\langle v - Ax, x' - x \rangle \geq 0 \quad \text{and} \quad - \langle v' - Ax', x' - x \rangle \geq 0.$$

Adding these two inequalities we obtain

$$\langle v - v' - A(x' - x), x' - x \rangle \geq 0,$$

hence, using (14),

$$\langle v - v', x' - x \rangle \geq \langle A(x' - x), x' - x \rangle \geq \alpha \|x - x'\|^2.$$

Taking norms and simplifying, we obtain that $A + N_C$ is strongly monotone with the constant α in (14).

Condition (14) is often called the *coercivity* condition, a name borrowed from physics, which is linked to preservation of energy represented by a quadratic form. We will use extensively the coercivity condition in the lectures about optimal control.

We will now show that for locally monotone mappings metric regularity and strong regularity come out to be the same.

Theorem 9.8. *If a mapping $F : H \rightrightarrows H$ that is locally monotone at $\bar{x}$ for $\bar{y}$ is metrically regular at $\bar{x}$ for $\bar{y}$, then it must be strongly regular at $\bar{x}$ for $\bar{y}$.*

Proofs

Proof of Reduction Lemma. Since we are only involved with local properties of C around one of its points $\bar{x}$, and $C - \bar{x}$ agrees with the cone $T_C(\bar{x})$ around 0 by Proposition 9.2, we can assume without loss of generality that $\bar{x} = 0$ and C is a cone, and $T_C(\bar{x}) = C$. Then, in terms of the polar cone C^*, which likewise is polyhedral, we have the characterization (4) in Proposition 9.1 that

$$v \in N_C(w) \iff w \in N_{C^*}(v) \iff w \in C, \quad v \in C^*, \quad \langle v, w\rangle = 0,$$

which gives us in particular that

$$N_{C^*}(\bar{v}) = \left\{ w \in C \,\middle|\, \langle \bar{v}, w\rangle = 0 \right\} = K. \tag{15}$$

We know on the other hand from Proposition 9.2 that

$$U \cap [C^* - \bar{v}] = U \cap T_{C^*}(\bar{v})$$

for some neighborhood U of 0, where $T_{C^*}(\bar{v})$ is polar to $N_{C^*}(\bar{v})$, hence equal to K^* by (15). Thus, there is a neighborhood O of $(0, 0)$ such that for $(w, u) \in O$ one has

$$\bar{v} + u \in N_C(w) \iff w \in C,\ u \in K^*,\ \langle \bar{v} + u, w\rangle = 0. \tag{16}$$

This may be compared with the fact that

$$u \in N_K(w) \iff w \in K,\ u \in K^*,\ \langle u, w\rangle = 0. \tag{17}$$

Our goal (in the context of $\bar{x} = 0$) is to show that (16) reduces to (9), at least when the neighborhood O in (16) is chosen smaller, if necessary. Because of (17), this comes down to demonstrating that $\langle \bar{v}, w\rangle = 0$.

We can take C to be represented by

$$C = \left\{ w \,\middle|\, \langle b_i, w\rangle \le 0 \ \text{ for } \ i = 1, \dots, m \right\},$$

in which case the polar C^* is represented by

$$C^* = \left\{ y_1 b_1 + \cdots + y_m b_m \,\middle|\, y_i \ge 0 \text{ for } i = 1, \dots, m \right\}. \tag{18}$$

Indeed, the polar to any cone K having a representation as in (2) consists of the vectors x satisfying $\langle b_i, x\rangle \le 0$ for $i = 1, \dots, m$. Inasmuch as $(K^*)^* = K$ for any closed convex cone K, we come to the representation (18). For each index set $I \subset \{1, \dots, m\}$, consider the polyhedral convex cones

$$W_I = \{ w \in C \,|\, \langle b_i, w\rangle = 0 \text{ for } i \in I\}, \qquad V_I = \{ \textstyle\sum_{i\in I} y_i b_i \text{ with } y_i \geq 0\},$$

with $W_\emptyset = C$ and $V_\emptyset = \{0\}$. Then $v \in N_C(w)$ if and only if, for some I, one has $w \in W_I$ and $v \in V_I$. In other words, gph N_C is the union of the finitely many polyhedral convex cones $G_I = W_I \times V_I$ in $I\!R^n \times I\!R^n$.

Among these cones G_I, we will only be concerned with the ones containing $(0, \bar{v})$. Let $\mathcal{I}$ be the collection of index sets $I \subset \{1, \ldots, m\}$ having that property. According to (7) in Proposition 9.2, there exists for each $I \in \mathcal{I}$ a neighborhood O_I of $(0, 0)$ such that $O_I \cap [G_I - (0, \bar{v})] = O_I \cap T_{G_I}(0, \bar{v})$. Furthermore, $T_{G_I}(0, \bar{v}) = W_I \times T_{V_I}(\bar{v})$. This yields that when $\bar{v} + u \in N_C(w)$ with (w, u) near enough to $(0, 0)$, we also have $\bar{v} + \tau u \in N_C(w)$ for all $\tau \in [0, 1]$. Since having $\bar{v} + \tau u \in N_C(w)$ entails having $\langle \bar{v} + \tau u, w\rangle = 0$, this implies that $\langle \bar{v}, w\rangle = -\tau\langle u, w\rangle$ for all $\tau \in [0, 1]$. Hence $\langle \bar{v}, w\rangle = 0$, as required. We merely have to shrink the neighborhood O in (16) to lie within every O_I for $I \in \mathcal{I}$. □

Proof of Theorem 9.8. According to Proposition 8.2, all we need to show is that a mapping F which is locally monotone and metrically regular at $\bar{x}$ for $\bar{y}$ must have a localization around $\bar{y}$ for $\bar{x}$ which is nowhere multivalued. Suppose to the contrary that every graphical localization of F^{-1} around $\bar{y}$ for $\bar{x}$ is multivalued. Then there are infinite sequences $y_k \to \bar{y}$ and $x_k, z_k \in F^{-1}(y_k)$, $x_k \to \bar{x}$, $z_k \to \bar{x}$ such that $x_k \neq z_k$ for all k. Let $b_k = \|z_k - x_k\| > 0$ and $h_k = (z_k - x_k)/b_k$. Then we have

$$\langle z_k, h_k\rangle = b_k + \langle x_k, h_k\rangle \text{ for } k = 1, 2, \ldots. \tag{19}$$

Since the metric regularity of F implies the Aubin property of F^{-1} at $\bar{y}$ for $\bar{x}$, there exist $\kappa > 0$ and $a > 0$ such that

$$F^{-1}(y) \cap I\!B_a(\bar{x}) \subset F^{-1}(y') + \kappa\|y - y'\|I\!B \quad \text{for all } y, y' \in I\!B_a(\bar{y}).$$

Choose a sequence of positive numbers τ_k satisfying $\tau_k \searrow 0$ and $\tau_k < b_k/2\kappa$. Then for k large, we have $y_k, y_k + \tau_k h_k \in I\!B_a(\bar{y})$ and $x_k \in F^{-1}(y_k) \cap I\!B_a(\bar{x})$, and hence there exists $u_k \in F^{-1}(y_k + \tau_k h_k)$ satisfying

$$\|u_k - x_k\| \leq \kappa\tau_k. \tag{20}$$

By the local monotonicity of F at $(\bar{x}, \bar{y})$ we have

$$\langle u_k - z_k, y_k + \tau_k h_k - y_k\rangle \geq 0.$$

This, combined with (19), yields

$$\langle u_k, h_k\rangle \geq \langle z_k, h_k\rangle \geq b_k + \langle x_k, h_k\rangle. \tag{21}$$

We get from (19), (20), and (21) that

$$b_k + \langle x_k, h_k \rangle \le \langle u_k, h_k \rangle \le \langle x_k, h_k \rangle + \kappa \tau_k < \langle x_k, h_k \rangle + (b_k/2),$$

which is impossible. Therefore, F^{-1} must have a localization around $\bar{y}$ for $\bar{x}$ which is single-valued. □

Lecture 10
Nonsmooth Inverse Function Theorems

The classical inverse function theorems assume continuous differentiability of the function involved. The more general Theorem 8.6 does not assume differentiability but considers a perturbation of the underlying mapping by a function with a sufficiently small Lipschitz modulus. In this lecture we present inverse function theorems in which the usual derivative is replaced by certain generalized derivatives of functions acting in finite dimensions. These are somewhat related to the generalized derivatives of set-valued mappings considered in Lecture 7, but we will go into this here.

According to a theorem by Rademacher, a function $f : I\!R^n \to I\!R^m$ that is Lipschitz continuous on an open set O is differentiable in a subset of O of full Lebesgue measure, and hence, in particular, differentiable at a set of points that is dense in O. If κ is a Lipschitz constant for f, then at those points of differentiability we have $\|\nabla f(x)\| \le \kappa$. Recall that a set $C \subset I\!R^n$ is said to be dense in a closed set D when the closure cl C of C coincides with D, or equivalently, when for any $x \in D$ any neighborhood U of x contains elements of C.

Consider now a function $f : I\!R^n \to I\!R^m$ and a point $\bar{x} \in \operatorname{int} \operatorname{dom} f$ where $\operatorname{lip}(f; \bar{x}) < \infty$. For any $\kappa > \operatorname{lip}(f; \bar{x})$ we have f Lipschitz continuous with constant κ in some neighborhood U of $\bar{x}$, and hence, from the Rademacher theorem, there is a dense set of points x in U where f is differentiable with $\|\nabla f(x)\| \le \kappa$. Hence there exist sequences $x_k \to \bar{x}$ such that f is differentiable at x_k, in which case the corresponding sequence of norms $\|\nabla f(x_k)\|$ is bounded by the Lipschitz constant κ and hence has at least one cluster point. This leads to the following definition of a set-valued derivative type concept:

Clarke Generalized Jacobian. *For $f : I\!R^n \to I\!R^m$ and any $\bar{x} \in \operatorname{int} \operatorname{dom} f$ where $\operatorname{lip}(f; \bar{x}) < \infty$, denote by $\bar{\nabla} f(\bar{x})$ the set consisting of all matrices $A \in I\!R^{m\times n}$ for which there is a sequence of points $x_k \to \bar{x}$ such that f is differentiable at x_k and $\nabla f(x_k) \to A$. The Clarke generalized Jacobian of f at $\bar{x}$, denoted $\bar{\partial} f(\bar{x})$, is the convex hull of this set: $\bar{\partial} f(\bar{x}) = \operatorname{co} \bar{\nabla} f(\bar{x})$.*

© The Author(s), under exclusive license to Springer Nature Switzerland AG 2021
A. L. Dontchev, *Lectures on Variational Analysis*, Applied Mathematical Sciences 205, https://doi.org/10.1007/978-3-030-79911-3_10

Note that $\bar{\nabla} f(\bar{x})$ is a nonempty, closed, and bounded subset of $\mathbb{R}^{m\times n}$. This ensures that the convex set $\bar{\partial} f(\bar{x})$ is nonempty, closed, and bounded as well. Furthermore, the mapping $x \mapsto \bar{\partial} f(x)$ has closed graph and is outer semicontinuous, meaning that, given $\bar{x} \in \operatorname{int} \operatorname{dom} f$, for every $\varepsilon > 0$ there exists $\delta > 0$ such that, for all $x \in \mathbb{B}_\delta(\bar{x})$,

$$\bar{\partial} f(x) \subset \bar{\partial} f(\bar{x}) + \varepsilon \mathbb{B}_{m\times n},$$

where $\mathbb{B}_{m\times n}$ is the set consisting of all $m \times n$ matrices whose norm is less or equal to one (the unit ball in the space of $m \times n$ matrices). Strict differentiability of f at $\bar{x}$ is known to be characterized by having $\bar{\nabla} f(\bar{x})$ consist of a single matrix A (or equivalently by having $\bar{\partial} f(\bar{x})$ consist of a single matrix A), in which case $A = \nabla f(\bar{x})$. We will use in this lecture a mean value theorem for the generalized Jacobian according to which, if f is Lipschitz continuous in an open convex set containing the points x and x', then one has

$$f(x) - f(x') = A(x - x') \quad \text{for some} \quad A \in \operatorname{co} \bigcup_{t\in[0,1]} \bar{\partial} f(tx + (1-t)x').$$

Clarke inverse function theorem stated below says roughly that a Lipschitz continuous function can be inverted locally around a point $\bar{x}$ when all elements of the generalized Jacobian at $\bar{x}$ are nonsingular matrices. Compared with the classical inverse function theorem, the main difference is that the function at hand is assumed Lipschitz continuous and not necessarily differentiable, and the single-valued graphical localization so obtained can only be claimed to be Lipschitz continuous.

Theorem 10.1 (Clarke). *Consider $f : \mathbb{R}^n \to \mathbb{R}^n$ and a point $\bar{x} \in \operatorname{int} \operatorname{dom} f$ where $\operatorname{lip}(f;\bar{x}) < \infty$. If all of the matrices in the generalized Jacobian $\bar{\partial} f(\bar{x})$ are nonsingular, then f is strongly regular at $\bar{x}$ for $\bar{y}$, meaning that f^{-1} has a Lipschitz continuous single-valued localization around $f(\bar{x})$ for $\bar{x}$.*

Let us illustrate the theorem by a couple of examples. The function $f : \mathbb{R} \to \mathbb{R}$ given by

$$f(x) = \begin{cases} x & \text{for } x < 0, \\ 2x & \text{for } x \geq 0 \end{cases}$$

has generalized Jacobian $\bar{\partial} f(0) = [1, 2]$, which does not contain 0. According to Clarke inverse function theorem, f^{-1} has a Lipschitz continuous single-valued localization around 0 for 0. In contrast, the function $f : \mathbb{R} \to \mathbb{R}$ given by $f(x) = |x|$ has $\bar{\partial} f(0) = [-1, 1]$, which does contain 0. Although the theorem makes no claims about this case, there is no graphical localization of f^{-1} around 0 for 0 that is single-valued.

We will deduce the Clarke theorem from an inverse function theorem due to B. Kummer, which relies on concept of strict graphical derivative defined as follows:

Strict Graphical Derivative. *For a function $f : \mathbb{R}^n \to \mathbb{R}^m$ and any point $\bar{x} \in \operatorname{dom} f$, the strict graphical derivative at $\bar{x}$ is the set-valued mapping $D_* f(\bar{x}) : \mathbb{R}^n \rightrightarrows \mathbb{R}^m$ defined by*

$$D_*f(\bar{x})(u) = \left\{ w \,\middle|\, w = \lim_{k\to\infty} \frac{f(x^k + \tau^k u^k) - f(x^k)}{\tau^k} \text{ for some } (x^k, u^k) \to (\bar{x}, u), \tau_k \searrow 0 \right\}.$$

When $\operatorname{lip}(f; \bar{x}) < \infty$, the set $D_*f(\bar{x})(u)$ is nonempty, closed, and bounded in $I\!R^m$ for each $u \in I\!R^n$. In this case the definition of $D_*f(\bar{x})(u)$ can be simplified by taking $u_k \equiv u$. In this setting the strict derivative can be expressed in terms of the generalized Jacobian. Specifically, it can be shown that $D_*f(\bar{x})(u) = \{ Au \mid A \in \bar{\partial} f(\bar{x}) \}$ for all u if $m = 1$, but that fails for higher dimensions. In general, for a function $f : I\!R^n \to I\!R^m$ with $\operatorname{lip}(f; \bar{x}) < \infty$, one has

$$\bar{\partial} f(\bar{x})(u) \supset D_*f(\bar{x})(u) \quad \text{for all} \quad u \in I\!R^n. \tag{1}$$

Note that f is strictly differentiable at $\bar{x}$ if and only if $D_*f(\bar{x})$ is a linear mapping, with the matrix for that mapping then being $\nabla f(\bar{x})$.

The computation of the strict derivatives will be illustrated next in an example.

Example. Consider the function

$$\theta^+ : x \mapsto x^+ := \max\{x, 0\}, \quad x \in I\!R.$$

Directly from the definition, we have, for any real u, that

$$D_*\theta^+(\bar{x})(u) = \begin{cases} u & \text{for } \bar{x} > 0, \\ \{ \lambda u \mid \lambda \in [0,1] \} & \text{for } \bar{x} = 0, \\ 0 & \text{for } \bar{x} < 0. \end{cases}$$

Similarly, the function

$$\theta^- : x \mapsto x^- := \min\{x, 0\}, \quad x \in I\!R$$

satisfies

$$\theta^-(x) = x - \theta^+(x).$$

Then, just by applying the definition, we get

$$v \in D_*\theta^+(\bar{x})(u) \iff (u - v) \in D_*\theta^-(\bar{x})(u) \text{ for any real } u.$$

We will now present a nonsmooth inverse function theorem due to B. Kummer, which furnishes a complete characterization of the strong regularity, and thus sharpens the Clarke theorem. This can also be deduced from the property in (1), which indicates that the Clarke inverse function theorem follows from that of Kummer.

Theorem 10.2 (Kummer). *Let $f : I\!R^n \to I\!R^n$ be continuous around $\bar{x}$. Then f^{-1} has a Lipschitz continuous single-valued localization around $f(\bar{x})$ for $\bar{x}$ if and only if*

$$0 \in D_*f(\bar{x})(u) \implies u = 0. \tag{2}$$

The proof of this theorem given below uses the following lemma:

Lemma 10.3. *For a function $f : I\!R^n \to I\!R^n$ that is continuous around $\bar{x}$, the inverse f^{-1} has a Lipschitz continuous single-valued localization around $f(\bar{x})$ for $\bar{x}$ if and only if there exist a neighborhood U of $\bar{x}$ and a constant $c > 0$ such that*

$$c\|x' - x\| \le \|f(x') - f(x)\| \quad \textit{for all } x', x \in U. \tag{3}$$

Proof. Let (3) hold. There is no loss of generality in supposing that U is open and f is continuous on U. For any $y \in f(U) := \{ f(x) \,|\, x \in U \}$, we have from (16) that if both $f(x) = y$ and $f(x') = y$ with x and x' in U, then $x = x'$; in other words, the mapping s which takes $y \in f(U)$ to $U \cap f^{-1}(y)$ is actually a function on $f(U)$. Moreover $\|s(y') - s(y)\| \le (1/c)\|y' - y\|$ by (3), so that this function is Lipschitz continuous relative to $f(U)$.

But this is not yet enough to get us to the desired conclusion about f^{-1}. For that, we need to know that s, or some restriction of s, is a localization of f^{-1} around $f(\bar{x})$ for $\bar{x}$, with graph equal to $(V \times U) \cap \operatorname{gph} f^{-1}$ for some neighborhood V of $f(\bar{x})$. The Brouwer invariance of domain theorem presented in the preceding Lecture 8 enters here: As applied to the restriction of f to the open set U, it tells us that $f(U)$ is open. We can therefore take $V = f(U)$ and the proof of the if direction is done.

Conversely, suppose that f^{-1} has a Lipschitz continuous single-valued localization around $f(\bar{x})$ for $\bar{x}$, its domain being a neighborhood V of $f(\bar{x})$. Let κ be a Lipschitz constant for s on V. Because f is continuous around $\bar{x}$, there is a neighborhood U of $\bar{x}$ such that $f(U) \subset V$. For x and x' in U, we have $s(f(x)) = x$ and $s(f(x')) = x'$, hence $\|x' - x\| \le \kappa\|f(x') - f(x)\|$. Thus, (16) holds for any $c > 0$ such that $c\kappa \le 1$. □

Proof of Theorem 10.2. We will show first that (2) implies (3), from which the sufficiency of the condition will follow. Assume there are sequences $c^k \to 0$, $x^k \to \bar{x}$, and $\tilde{x}^k \to \bar{x}$ such that

$$\|f(x^k) - f(\tilde{x}^k)\| \le c^k \|x^k - \tilde{x}^k\|.$$

Then the sequence of points

$$u^k := \frac{\tilde{x}^k - x^k}{\|x^k - \tilde{x}^k\|}$$

satisfies $\|u_k\| = 1$ for all k, hence a subsequence u^{k_i} of it is convergent to some $u \ne 0$. Restricting ourselves to such a subsequence, we obtain for $t^{k_i} = \|x^{k_i} - \tilde{x}^{k_i}\|$ that

$$\lim_{k_i \to \infty} \frac{f(x^{k_i} + t^{k_i} u^{k_i}) - f(x^{k_i})}{t^{k_i}} = 0.$$

By definition, the limit on the left side belongs to $D_* f(\bar{x})(u)$, yet $u \ne 0$, which is contrary to (2). Hence (2) does imply (3).

For the converse, we argue that if (2) were violated, there would be sequences $\tau^k \to 0$, $x^k \to \bar{x}$, and $u^k \to u$ with $u \ne 0$, for which

$$\lim_{k\to\infty} \frac{f(x^k + \tau^k u^k) - f(x^k)}{\tau^k} = 0. \tag{4}$$

On the other hand, under (3) one has

$$\|f(x^k + \tau^k u^k) - f(x^k)\|/\tau^k \geq c\|u^k\|,$$

which, combined with (4) and the assumption that u^k is away from 0 leads to a contradiction for large k. Thus (3) implies (2). □

In the following lecture we will apply Theorem 10.2 to a nonlinear programming problem dependent on a parameter.

We end this lecture by stating an extended version of the Clarke inverse function theorem for nonsmooth generalized equations.

Theorem 10.4. *Consider a function $f : I\!R^n \to I\!R^n$ and a set-valued mapping $F : I\!R^n \rightrightarrows I\!R^n$, as well as a point $(\bar{x}, \bar{y}) \in \operatorname{gph}(f + F)$ such that* $\operatorname{lip}(f; \bar{x}) < \infty$. *Assume that for each $A \in \bar{\partial} f(\bar{x})$ the mapping*

$$x \mapsto f(\bar{x}) + A(x - \bar{x}) + F(x)$$

is strongly regular at $\bar{x}$ for $\bar{y}$. Then the mapping $f + F$ is strongly regular at $\bar{x}$ for $\bar{y}$.

Compared with Corollary 8.7, Theorem 10.4 assumes less for the function f; namely, Lipschitz continuity instead of differentiability, but it also claims less. First, it is valid in finite dimensions and second, it gives only a sufficient condition for strong regularity of $f + F$ which is not necessary, in general.

Lecture 11
Lipschitz Stability in Optimization

In this lecture we consider optimization problems involving parameters with the aim to find conditions under which the corresponding solution mappings are Lipschitz continuous functions of the parameters. We start with the following quadratic optimization problem stated in a Hilbert space H:

$$\min \frac{1}{2}\langle x, Ax\rangle + \langle p, x\rangle \quad \text{subject to} \quad x \in C, \tag{1}$$

where $A : H \to H$ is a linear, bounded and self-adjoint mapping, C is a nonempty, closed, and convex set in H, and $p \in H$ is a parameter. From Lecture 1 we know that a necessary optimality condition for problem (1) is represented by the variational inequality

$$Ax + p + N_C(x) \ni 0, \tag{2}$$

where $N_C : H \rightrightarrows H$ is the normal cone mapping for the convex set C. In the final part of Lecture 9 we introduced the property of strong monotonicity and also showed that if the mapping $A + N_C$ satisfies the coercivity condition

$$\langle Ax, x\rangle \geq \alpha\|x\|^2 \quad \text{for every} \quad x \in C - C, \tag{3}$$

then it is strongly monotone. Condition (3) translates to problem (1) being strongly convex, which in turn implies that (2) is a necessary and sufficient optimality condition. We will add to that the claim that the solution mapping of (2) is globally a Lipschitz continuous function with respect to the parameter p:

Theorem 11.1. *Consider problem* (1) *and suppose that coercivity condition* (3) *holds with a constant* $\alpha > 0$. *Then problem* (1) *is equivalent to variational inequality* (2). *Furthermore, for each* $p \in H$ *problem* (1), *or equivalently* (2), *has a solution* $x(p)$ *which is unique and the function* $p \mapsto x(p)$ *is Lipschitz continuous on* H *with a Lipschitz constant* $1/\alpha$.

© The Author(s), under exclusive license to Springer Nature Switzerland AG 2021

A. L. Dontchev, *Lectures on Variational Analysis*, Applied Mathematical Sciences 205, https://doi.org/10.1007/978-3-030-79911-3_11

Proof. Denote by $v(p, x)$ the objective function of problem (1) and for any fixed $p \in H$. Choose any $\bar{x} \in C$ and consider the set

$$C(p) = \{x \in C \mid v(p, x) \le v(p, \bar{x})\}.$$

Observe that the set $C(p)$ is closed, convex, and nonempty. We will prove that this set is bounded. On the contrary, suppose that there is a sequence $x_k \in C(p)$ such that $\|x_k\| \to \infty$ as $k \to \infty$. Then we have

$$\frac{\alpha}{2}\|x_k\|^2 + \langle p, x_k\rangle \le \frac{1}{2}\langle x_k, Ax_k\rangle + \langle p, x_k\rangle \le v(p, \bar{x}).$$

Then

$$\frac{\alpha}{2}\|x_k\|^2 \le \|p\|\|x_k\| + v(p, \bar{x}),$$

hence

$$\alpha\|x_k\| \le 2\|p\| + \frac{2v(p, \bar{x})}{\|x_k\|}.$$

Passing to the limit in the last inequality, we obtain a contradiction.

As noted in preparatory lecture, any convex, closed, and bounded set in a Hilbert space is weakly compact. Furthermore, the quadratic objective function $v(p, \cdot)$ is convex and continuous, hence it is weakly lower semicontinuous. By the Weierstrass theorem, any problem of minimization of a weakly lower semicontinuous function over a weakly compact set has a solution. Thus, for each $p \in H$ problem (1) has a solution $x(p)$. Furthermore, the solution is unique, by the strict convexity of $v(p, \cdot)$. Let us now prove Lipschitz continuity of $x(p)$.

Let $p, p' \in H$ and let x, x', $x \ne x'$, be the corresponding unique solutions of (1), and hence also of (2). Then, by (2),

$$\langle Ax + p, x' - x\rangle \ge 0 \quad \text{and} \quad -\langle Ax' + p', x' - x\rangle \ge 0.$$

Adding these two inequalities we obtain

$$\langle A(x - x'), x' - x\rangle + \langle p - p', x' - x\rangle \ge 0.$$

Hence

$$\langle p - p', x' - x\rangle \ge \langle A(x' - x), x' - x\rangle.$$

Taking norms and using (3), we get

$$\|p - p'\|\|x - x'\| \ge \langle p - p', x' - x\rangle \ge \langle A(x' - x), x' - x\rangle \ge \alpha\|x' - x\|^2,$$

hence

$$\|x' - x\| \le \|p' - p\|/\alpha,$$

which completes the proof. □

Let us consider now problem (1) with the following constraint set which depends on a new parameter q:

$$C(q) = \{x \in H \mid Bx + q \in K\}, \tag{4}$$

where $q \in W$, W is a Hilbert space, $B : H \to W$ is a linear and bounded mapping, and K is a closed and convex cone in W. Then problem (1) takes the form

$$\min \frac{1}{2}\langle x, Ax\rangle + \langle p, x\rangle \quad \text{subject to} \quad Bx + q \in K. \tag{5}$$

Assume that the mapping B is surjective. Then condition (3) can be equivalently written as

$$\langle Ax, x\rangle \geq \|x\|^2 \quad \text{whenever} \quad Bx \in K - K. \tag{6}$$

This comes out from the following statement:

Exercise 11.2. *Show that if the mapping B is surjective, then*

$$C - C = \{x \in H \mid Bx \in K - K\}.$$

Guide. First show that the set $C - C$ is contained in $\{x \in H \mid Bx \in K - K\}$ by simply using the definitions of the sets. Then apply the surjectivity of B to show the opposite inclusion. □

The optimality condition for problem (5) will be derived in terms of Lagrange multipliers. In Lecture 1 we presented the Lagrange multiplier rule, or the Lagrange principle, for such a problem in finite dimensions. It is straightforward to extend it to Hilbert spaces. We obtain that under surjectivity of B, there exists a Lagrange multiplier y which is in the polar cone defined as

$$K^* = \{y \in H \mid \langle y, x\rangle \leq 0 \ \text{for all} \ x \in K\},$$

such that the following optimality system holds:

$$\begin{cases} Ax + B^*y + p = 0 \\ Bx + q + N_{K^*}(y) \ni 0 \end{cases} \tag{7}$$

We will consider now the dependence of the variables x and y on the joint parameter (p, q). Let $p_1, p_2 \in H, q_1, q_2 \in W$ and let x_1, x_2 be the corresponding unique solutions to problem (5). Remembering that by assumption the mapping B is surjective, by the comment after the Banach open mapping principle in Lecture 4, the mapping B^{-1} is Lipschitz continuous. Specifically, there exists a constant κ such that, given a solution $\tilde{x}_1$ of $Bx = -q_1$ there exists a solution $\tilde{x}_2$ of $Bx = -q_2$ such that $\|\tilde{x}_1 - \tilde{x}_2\| \leq \kappa\|q_1 - q_2\|$. Making the change of variables $x_i = w_i + \tilde{x}_i$ in (7) for $i = 1, 2$, we obtain

$$\begin{cases} Aw_i + B^*y_i + A\tilde{x}_i + p_i = 0 \\ Bw_i + N_{K^*}(y) \ni 0 \end{cases}$$

Hence, w_i, $i = 1, 2$ is the solution of

$$\min \frac{1}{2}\langle x, Ax\rangle + \langle A\tilde{x}_i + p_i, x\rangle \text{ subject to } Bx \in K.$$

Note that now the constraint does not depend on a parameter. By Theorem 10.1, it follows that

$$\alpha\|w_1 - w_2\| \le \|p_1 - p_2\| + \|A\|\|\tilde{x}_1 - \tilde{x}_2\| \le \|p_1 - p_2\| + \|A\|\kappa\|q_1 - q_2\|.$$

Using the equality $x_1 - x_2 = w_1 - w_2 + \tilde{x}_1 - \tilde{x}_2$, elementary rearrangements give us

$$\|x_1 - x_2\| \le c(\|p_1 - p_2\| + \|q_1 - q_2\|),$$

where the constant c is independent of the choice of $p_i, q_i, i = 1, 2$. Moreover, since the adjoint B^* is injective, we conclude that a similar estimate holds for the respective Lagrange multipliers $y_i, i = 1, 2$. We summarize the result obtained in the following theorem:

Theorem 11.3. *Consider problem* (5) *and suppose that the mapping B is surjective and condition* (6) *holds. Then for each $(p, q) \in H \times W$ the problem has a unique solution $x(p, q)$ and a unique Lagrange multiplier $y(p, q)$; moreover, the function $(p, q) \mapsto (x(p, q), y(p, q))$ is Lipschitz continuous on $H \times W$.*

We consider next the following nonlinear programming problem, with inequality constraints and special perturbations represented by the parameters v and u:

$$\text{minimize } g_0(x) - \langle v, x\rangle \text{ over all } x \text{ satisfying } g_i(x) \le u_i \text{ for } i \in [1, m], \tag{8}$$

where the functions $g_i : I\!R^n \to I\!R$, $i = 0, 1, \ldots, m$ are assumed to be twice continuously differentiable everywhere, for simplicity. According to the first-order optimality conditions displayed in Lecture 1, if x is a solution to (8) for some u and v and the Mangasarian–Fromovitz constraint qualification condition holds, then there exists a multiplier vector $y = (y_1, \ldots, y_m)$ such that the pair (x, y) satisfies the Karush–Kuhn–Tucker conditions, which are in the form of the variational inequality

$$\begin{pmatrix} -v + \nabla g_0(x) + y\nabla g(x) \\ -u + g(x), \end{pmatrix} \in N_E(x, y), \tag{9}$$

where $E = I\!R^n \times I\!R^m_+$.

We will now apply Theorem 10.2 to characterize the strong regularity of the mapping

$$G : (x, y) \mapsto \begin{pmatrix} \nabla g_0(x) + y\nabla g(x) \\ -g(x) \end{pmatrix} + N_E(x, y). \tag{10}$$

More specifically, we will give conditions under which the solution mapping of the inclusion

$$G(x, y) \ni \begin{pmatrix} v \\ u \end{pmatrix}$$

has a single-valued and Lipschitz continuous localization around the reference value $(\bar{v}, \bar{u})$ of the parameters for which $(\bar{x}, \bar{y})$ solves (9), that is, $(\bar{v}, \bar{u}) \in G(\bar{x}, \bar{y})$.

We first convert the variational inequality (9) into an equation involving the following function $H : I\!R^{n+m} \to I\!R^{n+m}$:

$$H(x, y) = \begin{pmatrix} \nabla g_0(x) + \sum_{i=1}^m y_i^+ \nabla g_i(x) \\ -g_1(x) + y_1^- \\ \vdots \\ -g_m(x) + y_m^- \end{pmatrix}, \tag{11}$$

where $y_i^+ = \max\{y_i, 0\}$ and $y_i^- = \min\{y_i, 0\}$. Let $((x, y), (v, u)) \in \operatorname{gph} H$; then for $z_i = y_i^+$, $i = 1, \dots, m$, we have that $((x, z), (v, u)) \in \operatorname{gph} G$. Indeed, for each $i = 1, \dots, m$, if $y_i \le 0$, then $u_i + g_i(x) = y_i^- \le 0$ and $(u_i + g_i(x))y_i^+ = 0$; otherwise $u_i + g_i(x) = y_i^- = 0$. Conversely, if $((x, z), (v, u)) \in \operatorname{gph} G$, then for

$$y_i = \begin{cases} z_i & \text{if } z_i > 0, \\ u_i + g_i(x) & \text{if } z_i = 0, \end{cases} \quad i = 1, \dots, m, \tag{12}$$

we obtain $(x, y) \in H^{-1}(v, u)$. In particular, if H^{-1} has a Lipschitz continuous localization around (v, u) for (x, y), then G^{-1} has the same property around (v, u) for (x, z) for $z_i = y_i^+$, $i = 1, \dots, m$, and if G^{-1} has a Lipschitz continuous localization around (v, u) for (x, z), then H^{-1} has the same property around (v, u) for (x, y), where y satisfies (12).

Next, we need to determine the strict graphical derivative of H. There is no trouble in differentiating the expressions $-g_i(x) + y_i^-$, inasmuch as we already know from the second example in the preceding lecture that the strict graphical derivative of y^+ is the negative of y^-. After some rearrangements and passing to the limit, we obtain

$$z \in D_*\varphi_i(\bar{x}, \bar{y})(u, v) \iff z = \bar{y}_i^+ \nabla^2 g_i(\bar{x})u + \lambda_i v_i \nabla g_i(\bar{x}), \quad i = 1, \dots, m,$$

where the coefficients λ_i for $i = 1, \dots, m$ satisfy

$$\lambda_i \begin{cases} = 1 & \text{for } \bar{y}_i > 0, \\ \in [0, 1] & \text{for } \bar{y}_i = 0, \\ = 0 & \text{for } \bar{y}_i < 0. \end{cases} \tag{13}$$

Directly from the definition, the strict graphical derivative of the function H in (11) at $(\bar{x}, \bar{y})$ satisfies

$$(\xi, \eta) \in D_*H(\bar{x}, \bar{y})(u, v) \iff \begin{cases} \xi = Au + \sum_{i=1}^m \lambda_i v_i \nabla g_i(\bar{x}), \\ \eta_i = -\nabla g_i(\bar{x})u + (1 - \lambda_i)v_i \ \text{ for } i = 1, \dots, m, \end{cases}$$

where the λ_i's are as in (13), and

$$A = \nabla^2 g_0(\bar{x}) + \sum \bar{y}_i^+ \nabla^2 g_i(\bar{x}). \tag{14}$$

Denoting by Λ the $m \times m$ diagonal matrix with elements λ_i on the diagonal, by I_m the $m \times m$ identity matrix, and setting

$$B = \begin{pmatrix} \nabla g_1(\bar{x}) \\ \vdots \\ \nabla g_m(\bar{x}) \end{pmatrix}, \tag{15}$$

we obtain that

$$M(\Lambda) \in D_* H(\bar{x}, \bar{y}) \iff M(\Lambda) = \begin{pmatrix} A & B^\top \Lambda \\ -B & I_m - \Lambda \end{pmatrix}. \tag{16}$$

This formula can be simplified by re-ordering the functions g_i according to the sign of $\bar{y}_i$.

Let $I = \{1, \ldots, m\}$ and, without loss of generality, suppose that

$$\{ i \in I \,|\, \bar{y}_i > 0 \} = \{1, \ldots, k\} \quad \text{and} \quad \{ i \in I \,|\, \bar{y}_i = 0 \} = \{k+1, \ldots, l\}.$$

Let

$$B_+ = \begin{pmatrix} \nabla g_1(\bar{x}) \\ \vdots \\ \nabla g_k(\bar{x}) \end{pmatrix} \quad \text{and} \quad B_0 = \begin{pmatrix} \nabla g_{k+1}(\bar{x}) \\ \vdots \\ \nabla g_l(\bar{x}) \end{pmatrix}, \tag{17}$$

let Λ_0 be the $(l-k) \times (l-k)$ diagonal matrix with diagonal elements $\lambda_i \in [0, 1]$, let I_0 be the identity matrix for $I\!R^{l-k}$, and let I_{m-l} be the identity matrix for $I\!R^{m-l}$. Then, since $\lambda_i = 1$ for $i = 1, \ldots, k$ and $\lambda_i = 0$ for $i = l+1, \ldots, m$, the matrix $M(\Lambda)$ in (16) takes the form

$$M(\Lambda_0) = \begin{pmatrix} A & B_+^\top & B_0^\top \Lambda_0 & 0 \\ 0 & 0 & 0 & 0 \\ -B & 0 & I_0 - \Lambda_0 & 0 \\ 0 & 0 & 0 & I_{m-l} \end{pmatrix}.$$

Each column of $M(\Lambda_0)$ depends on at most one λ_i, hence there are numbers

$$a_{k+1}, b_{k+1}, \ldots, a_l, b_l$$

such that

$$\det M(\Lambda_0) = (a_{k+1} + \lambda_{k+1} b_{k+1}) \cdots (a_l + \lambda_l b_l).$$

Therefore, $\det M(\Lambda_0) \neq 0$ for all $\lambda_i \in [0, 1]$, $i = k+1, \ldots, l$, if and only if the following condition holds:

$$a_i \neq 0,\ a_i + b_i \neq 0 \text{ and } [\text{sign } a_i = \text{sign}\,(a_i + b_i) \text{ or } \text{sign } b_i \neq \text{sign}\,(a_i + b_i)], \quad \text{for } i = k+1, \ldots, l.$$

Here we invoke the convention that

$$\text{sign } a = \begin{cases} 1 & \text{for } a > 0, \\ 0 & \text{for } a = 0, \\ -1 & \text{for } a < 0. \end{cases}$$

One can immediately note that it is not possible to have simultaneously sign $b_i \neq$ sign $(a_i + b_i)$ and sign $a_i \neq$ sign $(a_i + b_i)$ for some i. Therefore, it suffices to have

$$a_i \neq 0,\ a_i + b_i \neq 0,\ \text{sign } a_i = \text{sign}\,(a_i + b_i) \text{ for all } i = k+1, \ldots, l. \tag{18}$$

Now, let J be a subset of $\{k+1, \ldots, l\}$ and for $i = k+1, \ldots, l$, and let

$$\lambda_i^J = \begin{cases} 1 & \text{for } i \in J, \\ 0 & \text{otherwise.} \end{cases}$$

Let Λ^J be the diagonal matrix composed by these λ_i^J, and let $B_0(J) = \Lambda^J B_0$. Then we can write

$$M(J) := M(\Lambda^J) = \begin{pmatrix} A & B_+^\top & B_0(J)^\top & 0 \\ & 0 & 0 & 0 \\ -B & 0 & I_0^J & 0 \\ & 0 & 0 & I_l \end{pmatrix},$$

where I_0^J is the diagonal matrix having 0 as $(i-k)$-th element if $i \in J$ and 1 otherwise. Clearly, all the matrices $M(J)$ are obtained from $M(\Lambda_0)$ by taking each λ_i either 0 or 1. The condition (18) can be then written equivalently as

$$\det M(J) \neq 0 \text{ and sign } \det M(J) \text{ is the same for all } J. \tag{19}$$

Let the matrix $B(J)$ have as rows the row vectors $\nabla g_i(\bar{x})$ for $i \in J$. Re-ordering the last $m-k$ columns and rows of $M(J)$, if necessary, we obtain

$$M(J) = \begin{pmatrix} A & B_+^\top & B(J)^\top & 0 \\ -B & 0 & 0 & I \end{pmatrix},$$

where I is now the identity for $I\!R^{\{k+1,\ldots,m\}\setminus J}$. The particular form of the matrix $M(J)$ implies that $M(J)$ fulfills (19) if and only if (19) holds for just a part of it, namely, for the matrix

$$N(J) := \begin{pmatrix} A & B_+^\top & B(J)^\top \\ -B_+ & 0 & 0 \\ -B(J) & 0 & 0 \end{pmatrix}. \tag{20}$$

By applying Theorem 10.2, we finally arrive at the following result:

Theorem 11.4. *The mapping G defined in* (10) *which enters the KKT variational inequality* (9) *is strongly regular around $(\bar{v}, \bar{u})$ for $(\bar{x}, \bar{y})$ if and only if, for the matrix $N(J)$ in* (20), $\det N(J)$ *has the same nonzero sign for all $J \subset \{ i \in I \,|\, \bar{y}_i = 0 \}$.*

The reminder of this lecture is devoted to the following nonlinear programming problem with inequality and equality constraints and with a general parameterization:

$$\text{minimize}\, g_0(p,x) \text{ over all } x \text{ satisfying } g_i(p,x) \begin{cases} \leq 0 & \text{for } i \in [1,s], \\ = 0 & \text{for } i \in [s+1,m], \end{cases} \tag{21}$$

where the functions $g_0, g_1, \ldots, g_m$ are assumed to be twice continuously differentiable as acting from $\mathbb{R}^d \times \mathbb{R}^n$ to $\mathbb{R}$ and $p \in \mathbb{R}^d$ is a parameter. The associated Lagrange function has the form

$$L(p,x,y) = g_0(p,x) + y_1 g_1(p,x) + \cdots + y_m g_m(p,x)$$

and, under the Mangasarian–Fromovitz constraint qualification, the variational inequality capturing the associated first-order necessary optimality conditions becomes

$$f(p,x,y) + N_E(x,y) \ni (0,0), \tag{22}$$

where

$$\begin{cases} f(p,x,y) = (\nabla_x L(p,x,y), -\nabla_y L(p,x,y)), \\ E = \mathbb{R}^n \times [\mathbb{R}^s_+ \times \mathbb{R}^{m-s}]. \end{cases}$$

The pairs (x,y) satisfying this variational inequality are the Karush–Kuhn–Tucker (KKT) pairs for problem in (21). The Karush–Kuhn–Tucker solution mapping $S : \mathbb{R}^d \to \mathbb{R}^n \times \mathbb{R}^m$ is defined as

$$p \mapsto S(p) = \left\{ (x,y) \,\middle|\, f(p,x,y) + N_E(x,y) \ni (0,0) \right\}.$$

In the theorem given next, we present sufficient conditions for the existence of a Lipschitz continuous single-valued localization of the KKT solution mapping. First, introduce the notation

$$\begin{aligned} I &= \left\{ i \in [1,m] \,\middle|\, g_i(\bar{p},\bar{x}) = 0 \right\} \supset \{s+1,\ldots,m\}, \\ I_0 &= \left\{ i \in [1,s] \,\middle|\, g_i(\bar{p},\bar{x}) = 0 \text{ and } \bar{y}_i = 0 \right\} \subset I, \\ I_1 &= \left\{ i \in [1,s] \,\middle|\, g_i(\bar{p},\bar{x}) < 0 \right\}. \end{aligned}$$

Define the subspace

$$M^+ = \left\{ w \in \mathbb{R}^n \,\middle|\, w \perp \nabla_x g_i(\bar{p},\bar{x}) \text{ for all } i \in I \backslash I_0 \right\}.$$

Theorem 11.5. *For the KKT solution mapping S of problem* (21)*, let $(\bar{x},\bar{y}) \in S(\bar{p})$. Assume that the following conditions are both fulfilled:*

(a) *The gradients $\nabla_x g_i(\bar{p},\bar{x})$ for $i \in I$ are linearly independent,*

(b) $\langle w, \nabla^2_{xx} L(\bar{p},\bar{x},\bar{y})w \rangle > 0$ *for every nonzero $w \in M^+$.*

Then the mapping S has a Lipschitz continuous single-valued localization s around $\bar{p}$ for $(\bar{x},\bar{y})$, and this localization is semidifferentiable at $\bar{p}$ with semiderivative given by

$$Ds(\bar{p})(q) = \bar{s}(-Bq), \quad \textit{where } B = \nabla_p f(\bar{p}, \bar{x}, \bar{y}) = \begin{pmatrix} \nabla^2_{xp} L(\bar{p}, \bar{x}, \bar{y}) \\ -\nabla_p g_1(\bar{p}, \bar{x}) \\ \vdots \\ -\nabla_p g_m(\bar{p}, \bar{x}) \end{pmatrix}. \tag{23}$$

Moreover, for every p in some neighborhood of $\bar{p}$, the x component of $s(p)$ furnishes a local minimum in problem (21).

Proof. First, note that condition (a) implies the Mangasarian–Fromovitz constraint qualification and also condition (b) implies the second-order sufficient condition; it is actually stronger in that it takes into account only active constraints with positive Lagrange multipliers. Therefore sometimes condition (b) is called the *strong* second-order sufficient condition.

Without loss of generality, assume that all inequality constraints are active at the reference point, that is, $g_i(\bar{p}, \bar{x}) = 0$ for $i \in [1, s]$ which means that $I_1 = \emptyset$ and $l = m$. Indeed, small changes of p and x will not change the status of these constraints, and the corresponding Lagrange multipliers will remain zero; therefore, the conditions (a) and (b) in the theorem will not change. By applying Corollary 8.7, we reduce the problem to considering the linearized variational inequality

$$0 \in \begin{pmatrix} A & B^{\top} \\ B & 0 \end{pmatrix} \begin{pmatrix} x \\ y \end{pmatrix} + \begin{pmatrix} 0 \\ N_{\mathbb{R}^l_+}(y + \bar{y}) \end{pmatrix}, \tag{24}$$

where the matrices A and B are defined in (14) and (15) with $\nabla g_i(\bar{x})$ replaced by $\nabla_x g_i(\bar{p}, \bar{x})$, respectively, and, without loss of generality, the variables x and y are shifted from a neighborhood of $(\bar{x}, \bar{y})$ to a neighborhood of zero. Using the decomposition

$$B = \begin{pmatrix} B_+ \\ B_0 \end{pmatrix},$$

where B_+ and B_0 are defined as in (17) with $\nabla g_i(\bar{x})$ replaced by $\nabla_x g_i(\bar{p}, \bar{x})$, and noticing that, since $\bar{y}_i > 0$ for $i \in I \setminus I_0 = \{1, \ldots, k\}$, we have

$$N_{\mathbb{R}^l_+}(y + \bar{y}) = \begin{pmatrix} 0 \\ N_{\mathbb{R}^{l-k}_+}(y) \end{pmatrix}$$

the linear variational inequality (24) becomes

$$0 \in \begin{pmatrix} A & B_+^{\top} & B_0^{\top} \\ B_+ & 0 & 0 \\ B_0 & 0 & 0 \end{pmatrix} \begin{pmatrix} x \\ y_+ \\ y_0 \end{pmatrix} + \begin{pmatrix} 0 \\ 0 \\ N_{\mathbb{R}^{l-k}_+}(y_0) \end{pmatrix}. \tag{25}$$

Here, with some abuse of notation, we denote by y_+ the vector in $\mathbb{R}^k$ of the first k components of y and similarly for y_0. Note that from (b) the matrix A is positive definite over the cone K which now has the form

$$K = \begin{pmatrix} 0 \\ 0 \\ I\!R^{l-k}_{+} \end{pmatrix}$$

and hence

$$K - K = \begin{pmatrix} 0 \\ 0 \\ I\!R^{l-k} \end{pmatrix}.$$

Since $B_0 x \in I\!R^{l-k}$ always holds, we can reduce the coercivity condition (6) to

$$\langle Ax, x \rangle > 0 \quad \text{whenever} \quad B_+ x = 0,$$

which is exactly condition (b), because in finite dimensions (6) is equivalent to the same condition with > 0 instead of $\geq \|x\|^2$. Also note that the required surjectivity of the matrix B is exactly condition (a). Thus, we can apply Theorem 11.3 obtaining that the solution mapping of the variational inequality (25) has a Lipschitz continuous single-valued localization around $(\bar{x}, \bar{y})$ for 0. But then, according to Corollary 8.7, the solution mapping S of the nonlinear variational inequality (22) has a Lipschitz continuous single-valued localization around $\bar{p}$ for $(\bar{x}, \bar{y})$. The semidifferentiability of the localization and the formula (23) for the semiderivative follow from Theorem 9.5. It remains to observe that condition (b) implies the second-order sufficient condition for the perturbed problem (21) which in turn yields that under small enough perturbations the x component of the localization is a local minimum. □

Lecture 12
Strong Subregularity

"One-point" variants of the property of metric regularity can be obtained if in the definition we fix one of the points x or y at the reference values $\bar{x}$ or $\bar{y}$. Specifically, consider a mapping F acting between metric spaces and $(\bar{x}, \bar{y})$ in the graph of F. Then F is said to be *subregular* at $\bar{x}$ for $\bar{y}$ when there exists a neighborhood U of $\bar{x}$ and V of $\bar{y}$ such that

$$d(x, F^{-1}(\bar{y})) \leq \kappa d(\bar{y}, F(x) \cap V) \quad \text{for all} \quad x \in U. \tag{1}$$

If we fix instead x at $\bar{x}$ and leave y free, we obtain the so-called *semiregularity* property. It turns out that both subregularity and semiregularity do not enjoy, in general, stability under linearization, in contrast to metric regularity (Lyusternik–Graves theorem) and strong regularity (Robinson theorem).

We will now show that subregularity is implied by the outer Lipschitz continuity of the inverse F^{-1}, namely,

$$F^{-1}(y) \subset F^{-1}(\bar{y}) + \kappa \|y - \bar{y}\| \mathbb{B} \quad \text{for all} \quad y \in V. \tag{2}$$

Let (2) hold for a mapping $F : \mathbb{R}^n \rightrightarrows \mathbb{R}^m$ with $(\bar{x}, \bar{y}) \in \operatorname{gph} F$ with some κ and V. Then, for any neighborhood U of $\bar{x}$ we have that

$$F^{-1}(y) \cap U \subset F^{-1}(\bar{y}) + \kappa \|y - \bar{y}\| \mathbb{B} \quad \text{for all} \quad y \in V. \tag{3}$$

Let $x \in U$. If $F(x) \cap V = \emptyset$, then the right side of (1) is $+\infty$ and we are done. If not, observe that having $x \in U$ and $y \in F(x) \cap V$ is the same as having $x \in F^{-1}(y) \cap U$ and $y \in V$. Then from (3) we have that for such x and y the ball $x + \kappa\|y - \bar{y}\|\mathbb{B}$ should have a nonempty intersection with $F^{-1}(\bar{y})$, hence $d(x, F^{-1}(\bar{y})) \leq \kappa\|y - \bar{y}\|$. Taking infimum in this inequality with respect to $y \in F(x) \cap V$ gives us (1).

Consider an affine mapping $x \mapsto Ax + b$ for an $n \times m$ matrix A. This mapping is polyhedral, hence, according to Theorem 3.1, its inverse A^{-1} is Lipschitz continuous in its domain, hence it is outer Lipschitz continuous, and therefore it satisfies (2) for any $\bar{y}$ and any neighborhood V of $\bar{y}$. Thus, every affine mapping is subregular

© The Author(s), under exclusive license to Springer Nature Switzerland AG 2021

A. L. Dontchev, *Lectures on Variational Analysis*, Applied Mathematical Sciences 205, https://doi.org/10.1007/978-3-030-79911-3_12

at any $\bar{x}$ for any $\bar{y}$ with $(\bar{x}, \bar{y})$ in the graph of the mapping. Consider now a function $f : I\!R^n \to I\!R^m$ which is strictly differentiable at $\bar{x}$. The affine function $x \mapsto F(x) := f(\bar{x}) + \nabla f(\bar{x})(x - \bar{x})$ is subregular at $\bar{x}$ for $f(\bar{x})$. Furthermore, the function $x \mapsto g(x) = f(x) - f(\bar{x}) - \nabla f(\bar{x})(x - \bar{x})$ has a Lipschitz modulus at $\bar{x}$ equal to zero, and therefore, for any $l > 0$ there is a neighborhood of $\bar{x}$ such that g is Lipschitz continuous in U with Lipschitz constant l. We have that $g + F = f$, hence f is a perturbation of F in a neighborhood of $\bar{x}$ with a Lipschitz constant which could be arbitrarily small depending on the neighborhood. If subregularity were stable with respect to perturbations with sufficiently small Lipschitz constants, then f should be subregular at $\bar{x}$ for $f(\bar{x})$. But not every strictly differentiable function is subregular; for example, take $f(x) = x|x|$ for $x \in I\!R$ and $\bar{x} = 0$.

The instability of subregularity can be obviated by passing to a strengthened form of it called *strong subregularity*, whose definition is as follows:

Strong Subregularity. *A mapping $F : X \rightrightarrows Y$ acting between metric spaces X and Y is said to be strongly subregular at $\bar{x}$ for $\bar{y}$ if $(\bar{x}, \bar{y}) \in \text{gph } F$, and there is a constant $\kappa > 0$ along with neighborhoods U of $\bar{x}$ and V of $\bar{y}$ such that*

$$\rho(x, \bar{x}) \le \kappa d(\bar{y}, F(x) \cap V) \quad \textit{for all } x \in U. \tag{4}$$

The infimum of κ for which this holds is the modulus of strong subregularity denoted by $\text{subreg}\,(F; \bar{x}\,|\,\bar{y})$. *The absence of strong regularity is indicated by* $\text{subreg}\,(F; \bar{x}\,|\,\bar{y}) = +\infty$.

From this definition it follows that if a mapping F is strongly subregular at $\bar{x}$ for $\bar{y}$, then $\bar{x}$ is an isolated point in $F^{-1}(\bar{y})$ (Fig. 12.1). Indeed, suppose that there is a sequence of points $x_k \in F^{-1}(\bar{y})$, $x_k \ne \bar{x}$, convergent to $\bar{x}$. Then the right side of (4) for x_k will be zero for all sufficiently large k while the left side is not, a contradiction.

In Theorem 4.3 we established that metric regularity is equivalent to the Aubin property of the inverse. By definition, strong regularity is equivalent to the existence of a single-valued Lipschitz continuous localization of the inverse. Analogously, there is a property of a mapping that corresponds to strong subregularity of its inverse—this property is introduced next, where again X and Y are metric spaces with both metrics denoted by $\rho(\cdot, \cdot)$:

Isolated Calmness. *A mapping $S : Y \rightrightarrows X$ is said to have the isolated calmness property if there exist a constant $\kappa \ge 0$ and neighborhoods U of $\bar{x}$ and V of $\bar{y}$ such that*

$$\rho(x, \bar{x}) \le \kappa \rho(y, \bar{y}) \quad \textit{when } x \in S(y) \cap U \textit{ and } y \in V. \tag{5}$$

This can also be written as

$$S(y) \cap U \subset I\!B_{\kappa\rho(y,\bar{y})}(\bar{x}) \quad \textit{when } y \in V. \tag{6}$$

The following theorem shows the equivalence of strong subregularity with isolated calmness of the inverse:

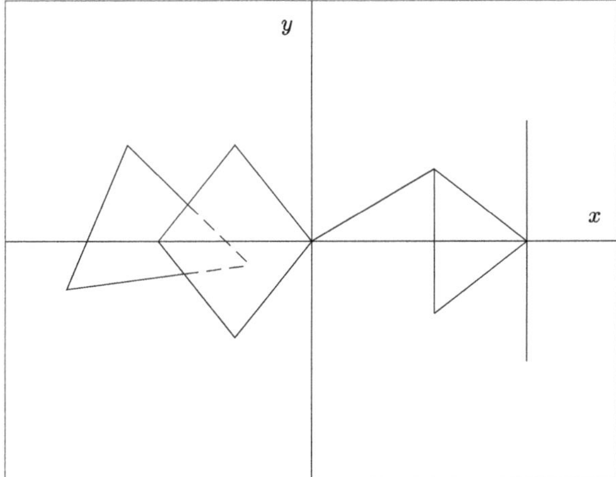

Fig. 12.1: Graph of a piecewise polyhedral mapping which is subregular at each point of its graph but is strongly subregular only at the origin

Theorem 12.1. *A mapping $F : X \rightrightarrows Y$ is strongly subregular at $\bar{x}$ for $\bar{y}$ with constant κ if and only if its inverse F^{-1} has the isolated calmness property at $\bar{y}$ for $\bar{x}$ with the same constant κ, i.e., there exist neighborhoods U of $\bar{x}$ and V of $\bar{y}$ such that*

$$F^{-1}(y) \cap U \subset I\!B_{\kappa\rho(y,\bar{y})}(\bar{x}) \quad \textit{for all} \quad y \in V. \tag{7}$$

Then infimum of all κ such that the inclusion holds for some U and V equals $\operatorname{subreg}(F; \bar{x} \,|\, \bar{y})$.

Proof. Assume first that F is strongly subregular at $\bar{x}$ for $\bar{y}$. Let $\kappa > \operatorname{subreg}(F; \bar{x} \,|\, \bar{y})$. Then there are neighborhoods U for $\bar{x}$ and V for $\bar{y}$ such that (4) holds with the indicated κ. Consider any $y \in V$. If $F^{-1}(y) \cap U = \emptyset$, then (7) holds trivially. If not, let $x \in F^{-1}(y) \cap U$. This entails $y \in F(x) \cap V$, hence $d(\bar{y}, F(x) \cap V) \le \rho(y, \bar{y})$ and consequently $\rho(x, \bar{x}) \le \kappa\rho(y, \bar{y})$ by (1); that is, (7) holds. Also, we see that $\operatorname{subreg}(F; \bar{x} \,|\, \bar{y})$ is not less than the infimum of all κ such that (7) holds for some choice of U and V.

For the converse, suppose (7) holds for some κ and neighborhoods U and V. Consider any $x \in U$. If $F(x) \cap V = \emptyset$ the right side of (4) is $+\infty$, and there is nothing more to prove. If not, for an arbitrary $y \in F(x) \cap V$ we have $x \in F^{-1}(y) \cap U$, and therefore, by (7), $\rho(x, \bar{x}) \le \kappa\rho(y, \bar{y})$. This being true for all $y \in F(x) \cap V$, we must have $\rho(x, \bar{x}) \le \kappa d(\bar{y}, F(x) \cap V)$. Thus, (4) holds, and in particular we have $\kappa \ge \operatorname{subreg}(F; \bar{x} \,|\, \bar{y})$. Thus, the infimum of κ in (7) equals $\operatorname{subreg}(F; \bar{x} \,|\, \bar{y})$. □

We look next at perturbations of the mapping F by a single-valued mapping g in the pattern that was followed for metric regularity and strong regularity. In particular, strong subregularity obeys a slightly modified form of the general paradigm as the other two regularity properties: *if a mapping F is strongly subregular, then for* *any function g with a sufficiently small calmness modulus, the mapping* $g + F$ *is strongly subregular as well.* Note that here, in contrast to metric regularity and strong regularity, the perturbation g is required to have a sufficiently small *calmness* modulus. This is established in the following theorem, which is an analogue of (Lyusternik–Graves) Theorem 5.2 and (Robinson) Theorem 8.6.

Theorem 12.2. *Let X be a metric space and Y be a linear metric space with a shift-invariant metric. Consider a mapping* $F : X \rightrightarrows Y$ *and a point* $(\bar{x}, \bar{y}) \in \operatorname{gph} F$ *such that F is strongly subregular at* $\bar{x}$ *for* $\bar{y}$*. Consider also a function* $g : X \to Y$ *with* $\bar{x} \in \operatorname{dom} g$*. Let* κ *and* μ *be nonnegative constants such that*

$$\kappa\mu < 1, \quad \operatorname{subreg}(F; \bar{x}\,|\,\bar{y}) \le \kappa \quad \textit{and} \quad \operatorname{clm}(g; \bar{x}) \le \mu.$$

Then

$$\operatorname{subreg}(g + F; \bar{x}\,|\,g(\bar{x}) + \bar{y}) \le \frac{\kappa}{1 - \kappa\mu}.$$

Proof. Choose κ and μ as in the statement of the theorem and let $\lambda > \kappa$, $\nu > \mu$ be such that $\lambda\nu < 1$. Pick $g : X \to Y$ with $\operatorname{clm}(g; \bar{x}) < \nu$; then there exists $a > 0$ such that

$$\rho(g(x), g(\bar{x})) \le \nu\rho(x, \bar{x}) \quad \text{when } x \in I\!B_a(\bar{x}). \tag{8}$$

Since $\operatorname{subreg}(F; \bar{x}\,|\,\bar{y}) < \lambda$, by taking a smaller if necessary, we have that

$$\rho(x, \bar{x}) \le \lambda\rho(y, \bar{y}) \quad \text{when } (x, y) \in \operatorname{gph} F \cap (I\!B_a(\bar{x}) \times I\!B_a(\bar{y})). \tag{9}$$

Let $\nu' = \max\{1, \nu\}$ and consider any

$$z \in I\!B_{a/2}(\bar{y}) \text{ with } x \in (g + F)^{-1}(z) \cap I\!B_{a/2\nu'}(\bar{x}). \tag{10}$$

Then $z \in g(x) + F(x)$, hence $z = y + g(x)$ for some $y \in F(x)$. From (8) and since $x \in I\!B_{a/2\nu'}(\bar{x})$, we have $\rho(g(x), g(\bar{x})) \le \nu(a/2\nu') \le a/2$ (inasmuch as $\nu' \ge \nu$). Using the equality $y - \bar{y} = z - g(x) - \bar{y}$, we get

$$\rho(y, \bar{y}) \le \rho(z, \bar{y}) + \rho(g(x), g(\bar{x})) \le (a/2) + (a/2) = a.$$

However, because $(x, y) \in \operatorname{gph} F \cap (I\!B_a(\bar{x}) \times I\!B_a(\bar{y}))$, through (9),

$$\rho(x, \bar{x}) \le \lambda\rho(z - g(x), \bar{y}) \le \lambda\rho(z, \bar{y}) + \lambda\rho(g(x), g(\bar{x})) \le \lambda\rho(z, \bar{y}) + \lambda\nu\rho(x, \bar{x}),$$

hence

$$\rho(x, \bar{x}) \le \lambda/(1 - \lambda\nu)\rho(z, \bar{y}).$$

Since x and z are chosen as in (10) and λ and ν could be arbitrarily close to κ and μ, respectively, the proof is complete. □

This last result implies in particular that the property of strong subregularity is preserved under linearization; compare with Corollaries 5.3 and 8.7 where it is required that the perturbation function is strictly Fréchet differentiable while here it is sufficient to assume (nonstrict) Fréchet

The above statement fails when the perturbation g is represented by a (calm) set-valued mapping even for $X = Y = I\!R$. Indeed, the mapping $F(x) = \{1 + x^2, 2x\}$ is strongly subregular at 0 for 0. Let $g(x) = \{-1, -x\}$; clearly g has the isolated calmness property at 0 for 0. However, as easily seen, the sum $g(x) + F(x) = \{x^2, 1 - x + x^2, 2x - 1, x\}$ is not strongly subregular at 0 for 0. Observe that the same counterexample shows that metric regularity is not preserved under a set-valued perturbation with a sufficiently small Lipschitz constant.

Corollary 12.3. *Let X and Y be Banach spaces. For a function $f : X \to Y$ which is differentiable at $\bar{x}$ and a set-valued mapping $F : X \rightrightarrows Y$ with $\bar{y} \in f(\bar{x}) + F(\bar{x})$,*

$$\operatorname{subreg}(f + F; \bar{x} \,|\, \bar{y}) \;=\; \operatorname{subreg}(f(\bar{x}) + Df(\bar{x})(\cdot - \bar{x}) + F(\cdot); \bar{x} \,|\, \bar{y}).$$

The following theorem furnishes an implicit function version of Theorem 12.2 and is analogous to Theorem 5.5 for metric regularity and Theorem 8.5 for strong regularity. Note that, in contrast to those theorems, the metric space X does not have to be complete.

Theorem 12.4. *Let P and X be metric spaces and let Y be a linear metric space with a shift-invariant metric. For a function $f : P \times X \to Y$ and a set-valued mapping $F : X \rightrightarrows Y$, consider the generalized equation*

$$f(p, x) + F(x) \ni 0$$

with solution mapping $p \mapsto S(p)$ having $\bar{x} \in S(\bar{p})$. Let κ and μ be positive constants such that $\kappa\mu < 1$. Let $h : X \to Y$ be a function which satisfies the conditions

$$h(\bar{x}) = f(\bar{p}, \bar{x}) \quad \text{and} \quad \widehat{\operatorname{clm}}_x(f - h; (\bar{p}, \bar{x})) \le \mu,$$

and suppose that $h + F$ is strongly subregular at $\bar{x}$ for 0 with

$$\operatorname{subreg}(h + F; \bar{x} \,|\, 0) \le \kappa.$$

Then the mapping S has the isolated calmness property at $\bar{p}$ for $\bar{x}$ with a constant κ' satisfying

$$\kappa' \le \frac{\kappa\lambda}{1 - \kappa\mu}.$$

A proof of Theorem 12.4 is given at the end of the lecture.

In the reminder of this lecture, we consider mappings acting between Euclidean spaces. This is motivated by the main application of the strong subregularity we will deal with, involving piecewise polyhedral mappings introduced in Lecture 3. Recall that a set-valued mapping is said to be piecewise polyhedral if its graph is the union of finitely many polyhedral sets. Also recall that a mapping $S : I\!R^m \rightrightarrows I\!R^n$ is outer Lipschitz continuous at $\bar{y} \in \operatorname{dom} S$ relative to $I\!R^n$ when $S(\bar{y})$ is closed and there is a constant $\kappa \ge 0$ along with a neighborhood V of $\bar{y}$ such that

$$S(y) \subset S(\bar{y}) + \kappa\|y - \bar{y}\|I\!B \quad \text{for all } \ y \in V. \tag{11}$$

According to Theorem 3.4, every piecewise polyhedral mapping is outer Lipschitz continuous. If in addition, $\bar{x}$ is an isolated point of $S(\bar{y})$, that is, $S(\bar{y}) \cap V = \{\bar{x}\}$, then from (11) we obtain

$$S(y) \subset \bar{x} + \kappa\|y - \bar{y}\|I\!B \quad \text{for all } \ y \in V,$$

which is equivalent to isolated calmness at $\bar{y}$ for $\bar{x}$. Combining these two observations gives us

Theorem 12.5. *A piecewise polyhedral mapping $F : I\!R^n \rightrightarrows I\!R^m$ is strongly subregular at $\bar{x}$ for $\bar{y}$ if and only if $\bar{x}$ is an isolated point of $F^{-1}(\bar{y})$, meaning that there is a neighborhood U of $\bar{x}$ such that $F^{-1}(\bar{y}) \cap U = \{\bar{x}\}$.*

Putting together Corollary 12.3 with Theorem 12.5, we obtain

Corollary 12.6. *Let $M : I\!R^n \rightrightarrows I\!R^m$ with $\bar{y} \in M(\bar{x})$ be of the form $M = f + F$ for $f : I\!R^n \to I\!R^m$ and $F : I\!R^n \rightrightarrows I\!R^m$ such that f is differentiable at $\bar{x}$ and F is polyhedral. Let $M_0(x) = f(\bar{x}) + \nabla f(\bar{x})(x - \bar{x}) + F(x)$. Then M^{-1} has the isolated calmness property at $\bar{y}$ for $\bar{x}$ if and only if $\bar{x}$ is an isolated point of $M_0^{-1}(\bar{y})$.*

In particular, a function $f : I\!R^n \to I\!R^m$ which is differentiable at $\bar{x}$ is strongly subregular at $\bar{x}$ for $f(\bar{x})$ if and only if its Jacobian $\nabla f(\bar{x})$ has rank n, so that $\nabla f(\bar{x})u = 0$ implies $u = 0$. It turns out that such a characterization can be provided also for set-valued mappings by letting graphical derivatives take over the role of ordinary derivatives. The following theorem, a proof of which is given at the end of the lecture, presents a graphical derivative criterion for strong subregularity.

Theorem 12.7. *A mapping $F : I\!R^n \rightrightarrows I\!R^m$ whose graph is locally closed at $(\bar{x}, \bar{y}) \in$ gph F is strongly subregular at $\bar{x}$ for $\bar{y}$ if and only if*

$$DF(\bar{x}\,|\,\bar{y})^{-1}(0) = \{0\}, \tag{12}$$

this being equivalent to

$$\|DF(\bar{x}\,|\,\bar{y})^{-1}\|^{+} < \infty, \tag{13}$$

and in that case

$$\operatorname{subreg}(F; \bar{x}\,|\,\bar{y}) = \|DF(\bar{x}\,|\,\bar{y})^{-1}\|^{+}. \tag{14}$$

Corollary 12.8. *Consider the variational inequality*

$$0 \in f(x) + N_C(x)$$

with a solution $\bar{x}$ and suppose that f is continuously differentiable around $\bar{x}$ and the convex set C is polyhedral. Let $A = \nabla f(\bar{x})$ and let $K = K_C(\bar{x}, \bar{v})$ be the corresponding critical cone for $\bar{v} = -f(\bar{x})$. Show that the condition

$$(A + N_K)^{-1}(0) = \{0\}, \tag{15}$$

is equivalent to the strong subregularity of the mapping $f + N_C$ at $\bar{x}$ for 0. *Also, in terms of the polar cone K^* to K, condition* (15) *is equivalent to*

$$w \in K, \quad -Aw \in K^*, \quad w \perp Aw \quad \Longrightarrow \quad w = 0. \tag{16}$$

Proof. To obtain (15), utilize (12) and the specific form of the graphical derivative established in Exercise 7.2. The equivalence of (15) and (16) follows from the relations between the normal cones to K and K^* given in Lecture 1 after the definition of the polar cone. □

As an application of Theorem 12.7, consider the minimization problem

$$\text{minimize} \;\; g(x) - \langle p, x\rangle \;\; \text{over} \;\; x \in C, \tag{17}$$

where C is a nonempty polyhedral subset of $I\!R^n$, $p \in I\!R^n$ is a parameter, and the function $g : I\!R^n \to I\!R$ is twice continuously differentiable everywhere. We first give a brief summary of the optimality conditions presented in Lecture 1.

If x is a local optimal solution of (17) for p, then x satisfies the necessary optimality condition

$$\nabla g(x) + N_C(x) \ni p. \tag{18}$$

Any solution x of (18) for p is a *stationary point* for problem (17) for p and the associated stationary point mapping is $p \mapsto S(p) = (\nabla g + N_C)^{-1}(p)$. The set of local minimizers of (17) for p is a subset of $S(p)$. If the function g is convex, then the set of local minimizers coincides with S(p); moreover, every stationary point is not only local but also a global minimizer. For the variational inequality (18), the critical cone to C associated with a solution x for p has the form

$$K_C(x, p - \nabla g(x)) = T_C(x) \cap [p - \nabla g(x)]^{\perp}.$$

If x furnishes a local minimum of (17) for p, then x must satisfy the second-order necessary condition

$$\langle u, \nabla^2 g(x)u\rangle \geq 0 \qquad \text{for all } \; u \in K_C(x, p - \nabla g(x)). \tag{19}$$

In addition, when $x \in S(p)$ satisfies the second-order sufficient condition

$$\langle u, \nabla^2 g(x)u\rangle > 0 \qquad \text{for all nonzero } \; u \in K_C(x, p - \nabla g(x)), \tag{20}$$

then x is a local optimal solution of (14) for p. Having x to satisfy (18) and (20) is equivalent to the existence of $\varepsilon > 0$ and $\delta > 0$ such that

$$g(y) - \langle p, y\rangle \geq g(x) - \langle p, x\rangle + \varepsilon\|y - x\|^2 \;\; \text{for all} \;\; y \in C \;\; \text{with} \;\; \|y - x\| \leq \delta, \tag{21}$$

meaning by definition that x furnishes a strong local minimum in (17).

Theorem 12.9. *Consider the stationary point mapping S for problem* (17), *that is, the solution mapping for* (18), *and let $\bar{x} \in S(\bar{p})$. Then the following are equivalent:*

(a) *The second-order sufficient condition* (20) *holds at* $\bar{x}$ *for* $\bar{p}$;

(b) *The point* $\bar{x}$ *is a local minimizer of* (17) *for* $\bar{p}$ *and the mapping* S *has the isolated calmness property at* $\bar{p}$ *for* $\bar{x}$.

Moreover, in either case, $\bar{x}$ *is actually a strong local minimizer for problem* (17) *for* $\bar{p}$.

Proof. Let $A := \nabla^2 g(\bar{x})$ and $K = K_C(\bar{x}, \bar{p} - \nabla g(\bar{x}))$. According to Corollary 12.8, the mapping S has the isolated calmness property at $\bar{p}$ for $\bar{x}$ if and only if condition (16) is satisfied. Let (a) hold. Then $\bar{x}$ is a local optimal solution as described. If (b) does not hold, there must exist some $u \neq 0$ satisfying the conditions in the left side of (16), and that would contradict the inequality $\langle u, Au \rangle > 0$ in (20).

Conversely, assume that (b) is satisfied. Then the second-order necessary condition (19) must hold; this can be written as

$$u \in K \quad \Longrightarrow \quad -Au \in K^*.$$

The isolated calmness property of S at $\bar{v}$ for $\bar{x}$ is characterized in (15), according to which there is no nonzero $u \in K$ such that the inequality in (16) fails to be strict. Thus, the necessary condition (19) turns into the sufficient condition (20). It remains to observe that (20) implies (21). $\square$

As a further application we study strong subregularity of the Karush–Kuhn–Tucker (KKT) mapping for the nonlinear programming problem:

$$\text{minimize } g_0(x) \text{ over all } x \text{ satisfying } g_i(x) \begin{cases} \leq 0 & \text{for } i \in [1, s], \\ = 0 & \text{for } i \in [s+1, m], \end{cases} \tag{22}$$

where the functions $g_i : I\!R^n \to I\!R$, $i = 0, 1, \ldots, m$ are twice continuously differentiable everywhere. Under the Mangasarian–Fromovitz constraint qualification condition, the first-order necessary optimality condition is represented by the KKT system:

$$\begin{cases} \nabla_x L(x, y) = 0, \\ \nabla_y L(x, y) \in N_{I\!R^s \times I\!R^{m-s}_+}, \end{cases} \tag{23}$$

where

$$L(x, y) = g_0(x) + \sum_{i=1}^{m} y_i g_i(x), \quad (x, y) \in I\!R^n \times I\!R^m,$$

is the Lagrange function associated with problem (22); here $y = (y_1, \ldots, y_m)$ is the vector of Lagrange multipliers. We study strong subregularity of the following mapping associated with the KKT system (23):

$$T : (x, y) \mapsto \begin{pmatrix} \nabla_x L(x, y) \\ -\nabla_y L(x, y) \end{pmatrix} + N_{I\!R^n \times I\!R^s \times I\!R^{m-s}_+}(x, y). \tag{24}$$

Let $(\bar{x}, \bar{y})$ be a reference solution of (23). Define the index sets

$$
\begin{aligned}
I_1 &= \{i \in \{s+1,\ldots,m\} \mid g_i(\bar{x}) = 0, \bar{y}_i > 0\} \cup \{1,\ldots,s\},\\
I_2 &= \{i \in \{s+1,\ldots,m\} \mid g_i(\bar{x}) = 0, \bar{y}_i = 0\},\\
I_3 &= \{i \in \{s+1,\ldots,m\} \mid g_i(\bar{x}) < 0, \bar{y}_i = 0\}.
\end{aligned}
$$

In further lines, we utilize the following condition:

$$
\begin{array}{l}
\text{there is no nonzero } y \in I\!R^m \text{ such that}\\
\sum_{i=1}^m y_i \nabla g_i(\bar{x}) = 0 \text{ and } y_i \geq 0,\ i \in I_2.
\end{array} \tag{25}
$$

This condition implies the Mangasarian–Fromovitz Constraint Qualification (MFCQ) condition, in which the set I_2 is replaced by $I_1 \cup I_2$. As well known, the MFCQ yields that the set of Lagrange multipliers for problem (22) satisfying (23) is nonempty, convex, and compact. Also recall that MFCQ is equivalent to the metric regularity of that mapping. In the proof of Theorem 12.10 given next, we will show that under (25) the set of Lagrange multipliers consists of a single point.

Denote $A = \nabla^2_{xx} L(\bar{x}, \bar{y})$ and $B = \nabla^2_{xy} L(\bar{x}, \bar{y})$; that is, B is the $n \times m$ matrix whose rows are the vectors $\nabla g_i(\bar{x}), i = 1, 2, \ldots, m$. Define the critical cone

$$
K = \{x' \mid \langle \nabla g_i(\bar{x}), x' \rangle = 0 \text{ for } i \in I_1,\ \langle \nabla g_i(\bar{x}), x' \rangle \leq 0 \text{ for } i \in I_2\}.
$$

Recall that the second-order necessary condition for local optimality has the form

$$
\langle x', Ax' \rangle \geq 0 \text{ for all } x' \in K,
$$

while the second-order sufficient condition is

$$
\langle x', Ax' \rangle > 0 \text{ for all } x' \in K \setminus \{0\}. \tag{26}
$$

Now we are ready to state and prove the following result:

Theorem 12.10. *The following are equivalent:*

(i) *The conditions* (25) *and* (26) *are both satisfied;*

(ii) *The KKT mapping T defined in (24) is strongly subregular at $(\bar{x}, \bar{y})$ for* 0, *and $\bar{x}$ is a strong local minimizer for* (22).

Proof. Linearizing the functions appearing in the mapping (24) at $(\bar{x}, \bar{y})$, we obtain the mapping

$$
L : (x, y) \mapsto \begin{pmatrix} 0 \\ \bar{g} \end{pmatrix} + \begin{pmatrix} A & B^T \\ -B & 0 \end{pmatrix} \begin{pmatrix} x - \bar{x} \\ y - \bar{y} \end{pmatrix} + N_{I\!R^n \times I\!R^s \times I\!R^{m-s}_+}(x, y), \tag{27}
$$

where we take into account that $\nabla_x L(\bar{x}, \bar{y}) = 0$ and $g_i(\bar{x}) = 0, i \in I_1 \cup I_2$, and use the notation

$$
\bar{g} = \begin{pmatrix} 0 \\ 0 \\ -g_{I_3}(\bar{x}) \end{pmatrix},
$$

in which g_I is a vector with components $g_i, i \in I$. We now apply Corollary 12.3 according to which the mapping T in (24) is strongly subregular at $(\bar{x}, \bar{y})$ for 0 if and only if the mapping L defined in (27) has the same property. The graph of the mapping L is the union of polyhedral sets hence the strong subregularity of T is equivalent to the property that the vector $(\bar{x}, \bar{y})$ is an isolated point in $L^{-1}(0)$.

Without loss of generality suppose that $I_1 = \{1, 2, \ldots, s_1\}$ and $I_2 = \{s_1 + 1, \ldots, s_2\}$. Denote by B_1 and B_2 the submatrices of B corresponding to the index sets I_1 and I_2, respectively; that is, the rows of B_1 are the vectors $\nabla g_i(\bar{x}), i = 1, 2, \ldots, s_1$, and analogously for B_2.

Let (i) hold. We will now show that (0, 0) is the unique solution of the variational inequality

$$\begin{aligned} &Ax + B^T y = 0, \\ &B_1 x = 0, \\ &B_2 x \in N_{\mathbb{R}_+^{I_2}}(y_{I_2}), \end{aligned} \tag{28}$$

where y_{I_2} is the subvector of y whose components have indices in I_2, and $\mathbb{R}_+^{I_2}$ is the set of vectors y_{I_2} with nonnegative components. Suppose that the mapping T is not strongly subregular at $(\bar{x}, \bar{y})$ for 0. Then there is a nonzero vector (x, y) satisfying (28). Assume that $x \neq 0$. Multiplying the first relation in (28) by x and taking into account the other two relations, we obtain $\langle x, Ax \rangle = 0$ which contradicts (26). Hence $x = 0$. But then there exists a nonzero $y \in \mathbb{R}^m$ such that $B^T y = 0$ and $0 \in N_{\mathbb{R}_+^{I_2}}(y_{I_2})$, hence $y_{I_2} \geq 0$. This contradicts (25). Thus the mapping T is strongly subregular at $(\bar{x}, \bar{y})$ for 0. It is standard that when $(\bar{x}, \bar{y})$ satisfies (23) and the second order sufficient condition (27) holds, then $\bar{x}$ is a strong local solution of problem (22). Hence, (ii) is established.

In the opposite direction, suppose that the conditions in (ii) are satisfied. Then from the analysis in the beginning of the proof, we conclude that the vector $(\bar{x}, \bar{y})$ as an isolated point in $L^{-1}(0)$. This in turn yields that $(0, 0)$ is the unique solution of the variational inequality (28). But this immediately implies (25). Furthermore, from the assumed optimality of $\bar{x}$ the second-order necessary condition holds:

$$\langle x', Ax' \rangle \geq 0 \text{ for all nonzero } x' \in K.$$

We only need to show that this inequality is strict. On the contrary, suppose that there exists a nonzero $x' \in K$ such that $Ax' = 0$. Then the nonzero vector $(x', 0)$ is a solution of (28), a contradiction. Hence the conditions in (i) are satisfied. □

Proofs

Proof of Theorem 12.4. Let κ, μ, and λ be as required and let $\delta > \kappa$ and $\nu > \mu$ be such that $\delta\nu < 1$. Let $\gamma > \lambda$. By the assumptions for the mapping $h + F$ and the functions f and h, there exist positive scalars a and r such that

$$\rho(x, \bar{x}) \leq \delta\rho(y, 0) \quad \text{for all } x \in (h + F)^{-1}(y) \cap I\!B_a(\bar{x}) \text{ and } y \in I\!B_{\nu a+\gamma r}(0), \tag{29}$$

$$\rho(f(p, x), f(\bar{p}, x)) \leq \gamma\rho(p, \bar{p}) \quad \text{for all } p \in I\!B_r(\bar{p}) \text{ and } x \in I\!B_a(\bar{x}), \tag{30}$$

and also, for $e = f - h$,

$$\rho(e(p, x), e(p, \bar{x})) \leq \nu\rho(x, \bar{x}) \quad \text{for all } x \in I\!B_a(\bar{x}) \text{ and } p \in I\!B_r(\bar{p}). \tag{31}$$

Let $x \in S(p) \cap I\!B_a(\bar{x})$ for some $p \in I\!B_r(\bar{p})$. Then, since $h(\bar{x}) = f(\bar{p}, \bar{x})$, we obtain from (30) and (31) that

$$\begin{aligned} \rho(e(p, x), 0) &\leq \rho(e(p, x), e(p, \bar{x})) + \rho(f(p, \bar{x}), f(\bar{p}, \bar{x})) \\ &\leq \nu\rho(x, \bar{x}) + \gamma\rho(p, \bar{p}) \leq \nu a + \gamma r. \end{aligned} \tag{32}$$

Observe that $x \in (h + F)^{-1}(-f(p, x) + h(x)) \cap I\!B_a(\bar{x})$, and then from (30) and (31) we have

$$\rho(x, \bar{x}) \leq \delta\rho(-f(p, x) + h(x), 0) \leq \delta\nu\rho(x, \bar{x}) + \delta\gamma\rho(p, \bar{p}).$$

In consequence,

$$\rho(x, \bar{x}) \leq \frac{\delta\gamma}{1 - \delta\nu}\rho(p, \bar{p}),$$

which gives us the desired result. □

Proof of Theorem 12.7. Suppose first that $\kappa > \operatorname{subreg}(F; \bar{x}\,|\,\bar{y})$ so that F is strongly subregular at $\bar{x}$ for $\bar{y}$ and (4) holds for some neighborhoods U and V. By definition, having $v \in DF(\bar{x}\,|\,\bar{y})(u)$ refers to the existence of sequences $u^k \to u$, $v^k \to v$, and $\tau^k \searrow 0$ such that $\bar{y} + \tau^k v^k \in F(\bar{x} + \tau^k u^k)$. Then $\bar{x} + \tau^k u^k \in U$ and $\bar{y} + \tau^k v^k \in V$ eventually, so that (4) yields $\|(\bar{x} + \tau^k u^k) - \bar{x}\| \leq \kappa\|(\bar{y} + \tau^k v^k) - \bar{y}\|$, which is the same as $\|u^k\| \leq \kappa\|v^k\|$. In the limit, this implies $\|u\| \leq \kappa\|v\|$. But then, by the characterization of the outer norm

$$\|H\|^{+} = \inf\Big\{ \kappa \in (0, \infty) \,\Big|\, y \in H(x) \to \|y\| \leq \kappa\|x\| \Big\} = \sup_{\|y\|=1} \frac{1}{d(0, H^{-1}(y))},$$

we have that $\|DF(\bar{x}\,|\,\bar{y})^{-1}\|^{+} \leq \kappa$ and hence

$$\operatorname{subreg}(F; \bar{x}\,|\,\bar{y}) \geq \|DF(\bar{x}\,|\,\bar{y})^{-1}\|^{+}. \tag{33}$$

In the other direction, (13) implies the existence of a $\kappa > 0$ such that

$$\sup_{v \in \mathbb{B}} \sup_{u \in DF(\bar{x}\,|\,\bar{y})^{-1}(v)} \|u\| < \kappa.$$

This in turn implies that $\|x - \bar{x}\| \leq \kappa\|y - \bar{y}\|$ for all $(x, y) \in \operatorname{gph} F$ close to $(\bar{x}, \bar{y})$. This gives us (4). Further, κ can be chosen arbitrarily close to $\|DF(\bar{x}\,|\,\bar{y})^{-1}\|^{+}$, and therefore

$$\|DF(\bar{x}\,|\,\bar{y})^{-1}\|^{+} \geq \operatorname{subreg}(F; \bar{x}\,|\,\bar{y}).$$

This, combined with (33), completes the proof. □

Lecture 13
Continuous Selections

The classical inverse function theorem presented in Lecture 8 gives conditions under which the inverse of a function has a single-valued localization, that is, locally, the inverse is a function. But what if every localization of the inverse is set-valued? In that case, could we claim the existence of a function that is "hidden" or "contained" in the inverse? To make things precise, we introduce the following concept:

Selections. *Given a set-valued mapping F acting between metric spaces X and Y and a set $D \subset \operatorname{dom} F$, a function s is said to be a selection. of F on D if $\operatorname{dom} s \supset D$ and $s(x) \in F(x)$ for all $x \in D$. If $(\bar{x}, \bar{y}) \in \operatorname{gph} F$, D is a neighborhood of $\bar{x}$ and s is a selection on D which satisfies $s(\bar{x}) = \bar{y}$, then s is said to be a local selection of F around $\bar{x}$ for $\bar{y}$.*

A selection of the inverse f^{-1} of a function $f : X \to Y$ might provide a *left inverse* or a *right inverse* to f. A left inverse to f on D is a selection $l : Y \to X$ of f^{-1} on $f(D)$ such that $l(f(x)) = x$ for all $x \in D$. Analogously, a right inverse to f on D is a selection $r : Y \to X$ of f^{-1} on $f(D)$ such that $f(r(y)) = y$ for all $y \in f(D)$. Commonly known are the right and the left inverses of the linear mapping from $I\!R^n$ to $I\!R^m$ represented by a matrix $A \in I\!R^{m\times n}$ that is of full rank. When $m \leq n$, the right inverse of the mapping corresponds to $A^{\mathsf{T}}(AA^{\mathsf{T}})^{-1}$, while when $m \geq n$, the left inverse corresponds to $(A^{\mathsf{T}}A)^{-1}A^{\mathsf{T}}$.

The following results show the existence of a local selection of the inverse of a function when the strict derivative is merely surjective.

Proposition 13.1. *Let X and Y be Hilbert spaces and let $f : X \to Y$ be a function which is strictly differentiable at $\bar{x}$ and such that the derivative $A := Df(\bar{x})$ is surjective. Then the inverse f^{-1} has a local selection s around $\bar{y} := f(\bar{x})$ for $\bar{x}$ which is strictly differentiable at $\bar{y}$ with derivative $Ds(\bar{y}) = A^*(AA^*)^{-1}$, where A^* is the adjoint of A.*

© The Author(s), under exclusive license to Springer Nature Switzerland AG 2021
A. L. Dontchev, *Lectures on Variational Analysis*, Applied Mathematical Sciences 205, https://doi.org/10.1007/978-3-030-79911-3_13

Proof. Consider the function

$$g : (x, u) \mapsto \begin{pmatrix} x + A^*u \\ f(x) \end{pmatrix} \quad \text{for } (x, u) \in X \times Y,$$

which satisfies $g(\bar{x}, 0) = (\bar{x}, \bar{y})$ and whose Jacobian is

$$J = \begin{pmatrix} I & A^* \\ A & 0 \end{pmatrix}.$$

In the Hilbert space context, if A is surjective then the operator J is invertible. Hence, by the classical inverse function theorem, the mapping g^{-1} has a single-valued graphical localization $(\xi, \eta) : (v, y) \mapsto (\xi(v, y), \eta(v, y))$ around $(\bar{x}, \bar{y})$ for $(\bar{x}, 0)$. In particular, for some neighborhoods U of $\bar{x}$ and V of $\bar{y}$, the function $s(y) := \xi(\bar{x}, y)$ satisfies $y = f(s(y))$ for $y \in V$. To obtain the formula for the strict derivative, use the inverse of J. □

The theorem of R. G. Bartle and L. M. Graves [5] given next extends Proposition 13.1 to Banach spaces:

Theorem 13.2 (Bartle–Graves). *Let X and Y be Banach spaces and let $f : X \to Y$ be a function which is strictly differentiable at $\bar{x}$ and such that the derivative $Df(\bar{x})$ is surjective. Then there is a neighborhood V of $\bar{y} := f(\bar{x})$ along with a continuous function $s : V \to X$ and a constant $\gamma > 0$ such that*

$$f(s(y)) = y \quad \textit{and} \quad \|s(y) - \bar{x}\| \le \gamma\|y - \bar{y}\| \quad \textit{for every} \ y \in V.$$

In other words, the surjectivity of the strict derivative at $\bar{x}$ implies that f^{-1} has a local selection s which is continuous around $f(\bar{x})$ and calm at $f(\bar{x})$. In Banach spaces the selection in the Bartle–Graves theorem, even for a bounded linear mapping, might be not linear. For this case we have:

Corollary 13.3. *For any mapping $A \in \mathcal{L}(X, Y)$, there is a continuous (but generally nonlinear) mapping B such that $ABy = y$ for every $y \in Y$.*

Recall that a mapping $F : X \rightrightarrows Y$ is (sequentially) inner semicontinuous on a set $T \subset X$ if for every $x \in T$, every $y \in F(x)$, and every sequence of points $x^k \in T$, $x^k \to x$, there exists $y^k \in F(x^k)$ for $k = 1, 2, \ldots$ such that $y^k \to y$ as $k \to \infty$. A basic result about selections is stated next:

Michael Selection Theorem. *Let X and Y be Banach spaces and consider a mapping $F : Y \rightrightarrows X$ which is closed and convex-valued and inner semicontinuous on* dom F. *Then F has a continuous selection $s :$* dom $F \to X$.

Note that the surjectivity of the derivative $Df(\bar{x})$ assumed in Theorem 13.2 is equivalent to metric regularity of f at $\bar{x}$ for $f(\bar{x})$. The following proposition extend that theorem to set-valued mappings by connecting metric regularity of a mapping with the existence of a continuous selection of its inverse. In what follows X and Y are Banach spaces.

Proposition 13.4. *Consider a set-valued mapping $F : X \rightrightarrows Y$ which is metrically regular at $\bar{x}$ for $\bar{y}$ and suppose that there exists a constant $c > 0$ such that the mapping $I\!B_c(\bar{y}) \ni y \mapsto F^{-1}(y) \cap I\!B_c(\bar{x})$ has convex values. Then for every $\alpha > \operatorname{reg}(F; \bar{x}\,|\,\bar{y})$ there exist a constant $\beta > 0$ and a continuous function $s : I\!B_\beta(\bar{y}) \to X$ such that*

$$s(y) \in F^{-1}(y) \quad \textit{and} \quad \|s(y) - \bar{x}\| \le \alpha d(y, F(\bar{x})) \quad \textit{for every } y \in I\!B_\beta(\bar{y}). \tag{1}$$

Proof. Choose α and κ such that $\alpha > \kappa > \operatorname{reg}(F; \bar{x}\,|\,\bar{y})$. Metric regularity of F at $\bar{x}$ for $\bar{y}$ means that there exist positive constants a and b such that for all $y \in I\!B_b(\bar{y})$,

$$F^{-1}(y) \cap I\!B_a(\bar{x}) \quad \text{is a closed set,} \tag{2}$$

and

$$d(x, F^{-1}(y)) \le \kappa d(y, F(x)) \quad \text{for all } x \in I\!B_a(\bar{x}). \tag{3}$$

Make a and b smaller if necessary so that $\min\{a, b\} < c$ and F^{-1} has the Aubin property at $\bar{y}$ for $\bar{x}$ with constants a, b and κ; that is,

$$F^{-1}(y) \cap I\!B_a(\bar{x}) \subset F^{-1}(y') + \kappa\|y - y'\| I\!B \quad \text{for all } y, y' \in I\!B_b(\bar{y}), \tag{4}$$

and also, by Proposition 4.1,

$$F^{-1}(y) \cap I\!B_a(\bar{x}) \neq \emptyset \quad \text{for all } y \in I\!B_b(\bar{y}). \tag{5}$$

Then choose a positive β such that

$$\beta < \min\left\{\frac{a}{\alpha}, b\right\} \tag{6}$$

and consider the mapping

$$I\!B_\beta(\bar{y}) \ni y \mapsto M(y) = \{x \mid x \in F^{-1}(y) \text{ and } \|x - \bar{x}\| \le \alpha d(y, F(\bar{x}))\}. \tag{7}$$

Let $y \in I\!B_\beta(\bar{y}))$. Then, from (3), since $\bar{y} \in F(\bar{x})$, and using (6), we have

$$d(\bar{x}, F^{-1}(y)) \le \kappa d(y, F(\bar{x})) \le \kappa\|y - \bar{y}\| \le \kappa\beta < \alpha\beta < a.$$

Thus $x \in F^{-1}(y)$ implies that $x \in F^{-1}(y) \cap I\!B_a(\bar{x})$, the latter set being closed by (2) and convex by the choice of $a < c$. We obtain that the set $M(y)$ is the intersection of a closed and convex set with a closed ball, hence it is closed and convex.

We show next that the mapping $I\!B_\beta(\bar{y}) \ni y \mapsto M(y)$ is inner semicontinuous. Let $(y, x) \in \operatorname{gph} M$. Then, using (6), we have

$$\|x - \bar{x}\| \le \alpha d(y, F(\bar{x})) \le \alpha\|y - \bar{y}\| \le \alpha\beta < a. \tag{8}$$

Choose any sequence $y_k \in I\!B_\beta(\bar{y})$, $y_k \to y$. We will find a sequence $x_k \to x$ such that $x_k \in M(y_k)$.

If $d(y_k, F(\bar{x})) = 0$, put $x_k = \bar{x}$. If $d(y_k, F(\bar{x})) > 0$, we pick x_k in the following way. From (3), since $y_k \in I\!B_b(\bar{y})$, we have from (3) that $d(\bar{x}, F^{-1}(y_k)) \le \kappa d(y_k, F(\bar{x}))$. Since $\kappa < \alpha$ and $d(y_k, F(\bar{x})) > 0$ we get $d(\bar{x}, F^{-1}(y_k)) < \alpha d(y_k, F(\bar{x}))$, hence there exists $\check{x}_k \in F^{-1}(y_k)$ such that

$$\|\check{x}_k - \bar{x}\| \le \alpha d(y_k, F(\bar{x})). \tag{9}$$

Note that

$$\|\check{x}_k - \bar{x}\| \le \alpha\|y_k - \bar{y}\| \le \alpha\beta < a < c. \tag{10}$$

From (8), $x \in F^{-1}(y) \cap I\!B_a(\bar{x})$. Recall that $y, y_k \in I\!B_b(\bar{y})$. Then, from the Aubin property (4) and (5) applied to x and y, y_k, we obtain that there exists $\hat{x}_k \in F^{-1}(y_k)$ such that

$$\|\hat{x}_k - x\| \le \kappa\|y_k - y\|. \tag{11}$$

Hence,

$$\begin{aligned}\|\hat{x}_k - \bar{x}\| \le \|x - \bar{x}\| + \kappa\|y_k - y\| &\le \alpha d(y, F(\bar{x})) + \kappa\|y_k - y\| \\ &\le \alpha d(y_k, F(\bar{x})) + (\alpha + \kappa)\|y_k - y\|.\end{aligned} \tag{12}$$

Taking into account (8) and that $a < c$, we have

$$\|\hat{x}_k - \bar{x}\| \le a + \kappa\|y_k - y\| < c,$$

for all sufficiently large k. Hence, using (10) and the last estimate, both $\check{x}_k$ and $\hat{x}_k$ are from $F^{-1}(y_k) \cap I\!B_c(\bar{x})$ for large k. Put

$$\varepsilon_k = \frac{(\alpha + \kappa)\|y_k - y\|}{(\alpha - \kappa)d(y_k, F(\bar{x})) + (\alpha + \kappa)\|y_k - y\|}.$$

Clearly $\varepsilon_k \in [0, 1]$ and $\varepsilon_k \to 0$ as $k \to \infty$. Let $x_k = \varepsilon_k\check{x}_k + (1 - \varepsilon_k)\hat{x}_k$. Then

$$x_k \in F^{-1}(y_k) \cap I\!B_c(\bar{x}) \subset F^{-1}(y_k).$$

Furthermore, using (9) and (12), we obtain

$$\|x_k - \bar{x}\| \le \varepsilon_k\kappa d(y_k, F(\bar{x})) + (1 - \varepsilon_k)(\alpha d(y_k, F(\bar{x})) + (\alpha + \kappa)\|y_k - \bar{x}\|) = \alpha d(y_k, F(\bar{x})),$$

hence $x_k \in M(y_k)$. Finally, from (11) and the fact that $\varepsilon_k \to 0$, we obtain $x_k \to x$. Thus, the mapping M defined in (7) is inner semicontinuous. By the Michael selection theorem, it has a continuous selection s. The definition of the map M yields that s satisfies (1). □

The following result extends both the Bartle–Graves Theorem 12.2 and Proposition 13.4. It shows that if a metrically regular mapping F, whose inverse is locally closed and convex, is perturbed by a function g with a sufficiently small Lipschitz modulus, then the inverse of the sum $g + F$ has a continuous and calm local selection. Its proof, as well as the proof of the following Theorem 13.6, are lengthy and technical, and are presented at the end of the lecture.

Theorem 13.5. *Consider a mapping $F : X \rightrightarrows Y$ and any $(\bar{x}, \bar{y}) \in \operatorname{gph} F$ and suppose that for some $c > 0$ the mapping $I\!B_c(\bar{y}) \ni y \mapsto F^{-1}(y) \cap I\!B_c(\bar{x})$ has convex values. Consider also a function $g : X \to Y$ with $\bar{x} \in \operatorname{int} \operatorname{dom} g$. Let κ and μ be nonnegative constants such that*

$$\kappa\mu < 1, \quad \operatorname{reg}(F; \bar{x} \,|\, \bar{y}) \le \kappa \quad \text{and} \quad \operatorname{lip}(g; \bar{x}) \le \mu.$$

Then for every γ with

$$\frac{\kappa}{1 - \kappa\mu} < \gamma$$

the mapping $(g + F)^{-1}$ has a continuous local selection s around $g(\bar{x}) + \bar{y}$ for $\bar{x}$, which moreover is calm at $g(\bar{x}) + \bar{y}$ with

$$\operatorname{clm}(s; g(\bar{x}) + \bar{y}) \le \gamma. \tag{13}$$

Proof of Theorem 13.2. Apply Theorem 13.5 with $F = Df(\bar{x})$ and $g(x) = f(x) - Df(\bar{x})x$. Metric regularity of F is equivalent to surjectivity of $Df(\bar{x})$, and F^{-1} is convex-closed-valued. The mapping g has $\operatorname{lip}(g; \bar{x}) = 0$ and finally $F + g = f$. □

The following theorem is an implicit mapping version of Theorem 13.5.

Theorem 13.6. *Let X, Y, and P be Banach spaces. For $f : P \times X \to Y$ and $F : X \rightrightarrows Y$, consider the generalized equation*

$$f(p, x) + F(x) \ni 0$$

with solution mapping $p \mapsto S(p)$ having $\bar{x} \in S(\bar{p})$. Suppose that F satisfies the assumptions of Theorem 13.5 with $\bar{y} = 0$ and associate constant $\kappa \ge \operatorname{reg}(F; \bar{x} \,|\, 0)$, and also that f is continuous on a neighborhood of $(\bar{p}, \bar{x})$ and has $\widehat{\operatorname{lip}}_x(f; (\bar{p}, \bar{x})) \le \mu$, where μ is a nonnegative constant satisfying $\kappa\mu < 1$. Then for every γ satisfying

$$\frac{\kappa}{1 - \kappa\mu} < \gamma,$$

there exist neighborhoods U of $\bar{x}$ and Q of $\bar{p}$ along with a continuous function $s : Q \to U$ such that

$$s(p) \in S(p) \quad \text{and} \quad \|s(p) - \bar{x}\| \le \gamma \|f(p, \bar{x}) - f(\bar{p}, \bar{x})\| \quad \text{for every} \quad p \in Q.$$

In the last theorem in this lecture, Theorem 13.7, given next we assume less than in Theorem 13.6 and obtain less. First, note that the theorem is stated in finite dimensions, mainly because its proof relies on Brouwer fixed point theorem given below. Most importantly, the condition for the Lipschitz modulus of the perturbation function is replaced by a condition for the calmness modulus. The conclusion, however, is that there exists a nonempty-valued selection which is continuous at $\bar{p}$ but may not be continuous around $\bar{p}$.

Brouwer Fixed Point Theorem. *Let Q be a compact and convex set in $I\!R^n$, and let $\Phi : I\!R^n \to I\!R^n$ be a function which is continuous on Q and maps Q into itself. Then there exists a point $x \in Q$ such that $\Phi(x) = x$.*

Theorem 13.7. *Consider the generalized equation*

$$0 \in f(p, x) + F(x) \tag{14}$$

with solution mapping $p \mapsto S(p)$, where $f : I\!R^d \times I\!R^m \to I\!R^n$ is a function of a parameter p and solution variable x, and $F : I\!R^m \rightrightarrows I\!R^n$ is a set-valued mapping. Suppose that (14) *has a reference solution $\bar{x}$ for the value $\bar{p}$ of the parameter; that is, $\bar{x} \in S(\bar{p})$. Also, let F be metrically regular at $\bar{x}$ for $\bar{y} := -f(\bar{p}, \bar{x})$ and let there exist a constant $c > 0$ such that the mapping $I\!B_c(\bar{y}) \ni y \mapsto F^{-1}(y) \cap I\!B_c(\bar{x})$ has convex values. Let $(\bar{p}, \bar{x}) \in \operatorname{int} \operatorname{dom} f$, and let f be continuous around $(\bar{p}, \bar{x})$ and such that*

$$\operatorname{reg}(F; \bar{x} \,|\, \bar{y}) \cdot \widehat{\operatorname{clm}}_x(f; (\bar{p}, \bar{x})) < 1. \tag{15}$$

Then, for every γ satisfying

$$\frac{\operatorname{reg}(F; \bar{x} \,|\, \bar{y})}{1 - \operatorname{reg}(F; \bar{x} \,|\, \bar{y}) \cdot \widehat{\operatorname{clm}}_x(f; (\bar{p}, \bar{x}))} < \gamma, \tag{16}$$

there exists a neighborhood Q of $\bar{p}$ such that for every $p \in Q$ there exists a solution $\sigma(p)$ of (14) *which satisfies the estimate*

$$\|\sigma(p) - \bar{x}\| \le \gamma d(0, f(p, \bar{x}) + F(\bar{x})). \tag{17}$$

In particular, the solution mapping S has a local selection σ around $\bar{p}$ for $\bar{x}$ which is continuous at $\bar{p}$. If, in addition, the function $f(\cdot, \bar{x})$ is calm at $\bar{p}$, then the selection σ is calm at $\bar{p}$ as well.

Proofs

Proof of Theorem 13.5. The proof consists of two steps. In the first step, we use induction to obtain a Cauchy sequence of continuous functions $z^0, z^1, \ldots$, such that z^n is a continuous and calm selection of the mapping $y \mapsto F^{-1}(y - g(z^{n-1}(y)))$. Then we show that this sequence has a limit in the space of continuous functions acting from a fixed ball around $\bar{y}$ to the space X, and this limit is the selection whose existence is claimed.

Choose κ and μ as in the statement of the theorem and let $\gamma > \kappa/(1 - \kappa\mu)$. Let λ, α, and ν be such that $\kappa < \lambda < \alpha < 1/\nu$ and $\nu > \mu$, and also $\lambda/(1 - \alpha\nu) \le \gamma$. Without loss of generality, we can assume that $g(\bar{x}) = 0$. Let $I\!B_a(\bar{x})$ and $I\!B_b(\bar{y})$ be the neighborhoods of $\bar{x}$ and $\bar{y}$, respectively, that are associated with the assumed properties of the mapping F and the function g. Specifically,

(a) For every $y, y' \in \mathbb{B}_b(\bar{y})$ and $x \in F^{-1}(y) \cap \mathbb{B}_a(\bar{x})$ there exists $x' \in F^{-1}(y')$ with

$$\|x' - x\| \le \lambda \|y' - y\|.$$

(b) For every $y \in \mathbb{B}_b(\bar{y})$ the set $F^{-1}(y) \cap \mathbb{B}_a(\bar{x})$ is nonempty, closed, and convex.

(c) The function g is Lipschitz continuous on $\mathbb{B}_a(\bar{x})$ with a constant ν.

According to Proposition 13.4, we can find a constant β, $0 < \beta \le b$, and a continuous function $z^0 : \mathbb{B}_\beta(\bar{y}) \to X$ such that

$$F(z^0(y)) \ni y \quad \text{and} \quad \|z^0(y) - \bar{x}\| \le \lambda \|y - \bar{y}\| \quad \text{for all } y \in \mathbb{B}_\beta(\bar{y}).$$

Choose a positive τ such that

$$\tau \le (1 - \alpha\nu) \min\left\{a, \frac{a}{2\lambda}, \frac{\beta}{2}\right\} \tag{18}$$

and consider the mapping $y \mapsto M_1(y)$ defined as

$$M_1(y) = \left\{ x \in F^{-1}(y - g(z^0(y))) \,\middle|\, \|x - z^0(y)\| \le \alpha\nu \|z^0(y) - \bar{x}\| \right\}$$

for $y \in \mathbb{B}_\tau(\bar{y})$ and $M_1(y) = \emptyset$ for $y \notin \mathbb{B}_\tau(\bar{y})$. Clearly, $(\bar{y}, \bar{x}) \in \operatorname{gph} M_1$. Also, for any $y \in \mathbb{B}_\tau(\bar{y})$, we have from the choice of λ and α, using (14), that $z^0(y) \in \mathbb{B}_a(\bar{x})$ and therefore

$$\begin{aligned}
\|y - g(z^0(y)) - \bar{y}\| &\le \|y - \bar{y}\| + \|g(z^0(y)) - g(\bar{x}\| \\
&\le \tau + \nu \|z^0(y) - \bar{x}\| \le \tau + \nu\lambda\tau \\
&\le (1 - \alpha\nu)(1 + \nu\lambda)(\beta/2) \le \beta \le b.
\end{aligned}$$

Then from the Aubin property of F^{-1}, there exists $x \in F^{-1}(y - g(z^0(y)))$ with

$$\|x - z^0(y)\| \le \lambda \|g(z^0(y)) - g(\bar{x})\| \le \alpha\nu \|z^0(y) - \bar{x}\|,$$

which implies $x \in M_1(y)$. Thus M_1 is nonempty-valued. Further, if $(y, x) \in \operatorname{gph} M_1$, using (18) we have that $y \in \mathbb{B}_b(\bar{y})$ and also

$$\|x - \bar{x}\| \le \|x - z^0(y)\| + \|z^0(y) - \bar{x}\| \le (1 + \alpha\nu)\lambda\tau \le (1 - (\alpha\nu)^2)\lambda \frac{a}{2\lambda} \le \frac{a}{2}.$$

Then, from the property (b) above, since for any $y \in \operatorname{dom} M_1$ the set $M_1(y)$ is the intersection of a closed ball with a closed convex set, the mapping M_1 is closed-convex-valued on its domain. We will show that this mapping is inner semicontinuous on $\mathbb{B}_\tau(\bar{y})$.

Let $y \in \mathbb{B}_\tau(\bar{y})$ and $x \in M_1(y)$, and let $y^k \in \mathbb{B}_\tau(\bar{y})$, $y^k \to y$ as $k \to \infty$. If $z^0(y) = \bar{x}$, then $M_1(y) = \{\bar{x}\}$ and therefore $x = \bar{x}$. Any $x^k \in M_1(y^k)$ satisfies

$$\|x^k - z^0(y^k)\| \le \alpha\nu \|z^0(y^k) - \bar{x}\|.$$

Using the continuity of the function z^0, we see that $x^k \to z^0(y) = \bar{x} = x$; thus M_1 is inner semicontinuous in this case.

Now let $z^0(y) \neq \bar{x}$. Since $z^0(y^k) \in F^{-1}(y^k - g(\bar{x})) \cap I\!B_a(\bar{x})$, the Aubin property of F^{-1} furnishes the existence of $\check{x}^k \in F^{-1}(y^k - g(z^0(y^k)))$ such that

$$\|\check{x}^k - z^0(y^k)\| \le \lambda \|g(z^0(y^k)) - g(\bar{x})\| \le \lambda \nu \|z^0(y^k) - \bar{x}\| \le \alpha \nu \|z^0(y^k) - \bar{x}\|, \quad (19)$$

where we also use the Lipschitz continuity of g. Then $\check{x}^k \in M_1(y^k)$, and in particular, $\check{x}^k \in I\!B_a(\bar{x})$. Further, the inclusion $x \in F^{-1}(y - g(z^0(y))) \cap I\!B_a(\bar{x})$ combined with the Aubin property of F^{-1} entails the existence of $\tilde{x}^k \in F^{-1}(y^k - g(z^0(y^k)))$ such that

$$\|\tilde{x}^k - x\| \le \lambda(\|y^k - y\| + \nu \|z^0(y^k) - z^0(y)\|) \to 0 \quad \text{as} \quad k \to \infty. \quad (20)$$

Then $\tilde{x}^k \in I\!B_a(\bar{x})$ for large k. Let

$$\varepsilon^k := \frac{(1 + \alpha\nu)\|z^0(y^k) - z^0(y)\| + \|\tilde{x}^k - x\|}{\alpha\nu \|z^0(y) - \bar{x}\| - \lambda\nu \|z^0(y^k) - \bar{x}\|}.$$

Note that, for $k \to \infty$, the numerator in the definition of ε^k goes to 0 because of the continuity of z^0 and (7), while the denominator converges to $(\alpha - \lambda)\nu\|z^0(y) - \bar{x}\| > 0$; therefore $\varepsilon^k \to 0$ as $k \to \infty$, and hence $\varepsilon_k \in (0, 1)$ for large k. Let

$$x^k = \varepsilon^k \check{x}^k + (1 - \varepsilon^k)\tilde{x}^k.$$

Since $\tilde{x}^k \to x$ and $\varepsilon^k \to 0$, we get $x^k \to x$ as $k \to \infty$ and also, since $y \mapsto F^{-1}(y) \cap I\!B_a(\bar{x})$ is convex-valued around $(\bar{x}, \bar{y})$, we have $x^k \in F^{-1}(y^k - g(z^0(y^k)))$ for large k. By (19), (20), the assumption that $x \in M_1(y)$, and the choice of ε^k, we have

$$\begin{aligned}
\|x^k - z^0(y^k)\| &\le \varepsilon^k \|\check{x}^k - z^0(y^k)\| + (1 - \varepsilon^k)\|\tilde{x}^k - z^0(y^k)\| \\
&\le \varepsilon^k \lambda\nu \|z^0(y^k) - \bar{x}\| + (1 - \varepsilon^k)(\|\tilde{x}^k - x\| \\
&\quad + \|x - z^0(y)\| + \|z^0(y) - z^0(y^k)\|) \\
&\le \varepsilon^k \lambda\nu \|z^0(y^k) - \bar{x}\| + \|\tilde{x}^k - x\| \\
&\quad + (1 - \varepsilon^k)\alpha\nu \|z^0(y) - \bar{x}\| + \|z^0(y) - z^0(y^k)\| \\
&\le \alpha\nu \|z^0(y^k) - \bar{x}\| + \alpha\nu \|z^0(y^k) - z^0(y)\| \\
&\quad + \|\tilde{x}^k - x\| + \|z^0(y) - z^0(y^k)\| \\
&\quad - \varepsilon^k \alpha\nu \|z^0(y) - \bar{x}\| + \varepsilon^k \lambda\nu \|z^0(y^k) - \bar{x}\| \\
&\le \alpha\nu \|z^0(y^k) - \bar{x}\| + \|\tilde{x}^k - x\| + (1 + \alpha\nu)\|z^0(y) - z^0(y^k)\| \\
&\quad - \varepsilon^k(\alpha\nu \|z^0(y) - \bar{x}\| - \lambda\nu \|z^0(y^k) - \bar{x}\|) \\
&= \alpha\nu \|z^0(y^k) - \bar{x}\|.
\end{aligned}$$

We obtain that $x^k \in M_1(y^k)$, and since $x^k \to x$, we conclude that the mapping M_1 is inner semicontinuous on its domain $I\!B_\tau(\bar{y})$. Hence, by Michael selection theorem, it has a continuous selection $z^1 : I\!B_\tau(\bar{y}) \to X$; that is, a continuous function z^1 which satisfies

$$z^1(y) \in F^{-1}(y - g(z^0(y))) \text{ and } \|z^1(y) - z^0(y)\| \le \alpha\nu\|z^0(y) - \bar{x}\| \text{ for all } y \in I\!B_\tau(\bar{y}).$$

Then for $y \in I\!B_\tau(\bar{y})$, by the choice of γ,

$$\|z^1(y) - \bar{x}\| \le \|z^1(y) - z^0(y)\| + \|z^0(y) - \bar{x}\| \le (1 + \alpha\nu)\lambda\|y - \bar{y}\| \le \gamma\|y - \bar{y}\|.$$

The induction step is parallel to the first step. Let z^0 and z^1 be as above and suppose we have also found functions $z^1, z^2, \ldots, z^n$, such that each $z^j, j = 1, 2, \ldots, n$, is a continuous selection of the mapping $y \mapsto M_j(y)$, where

$$M_j(y) = \left\{ x \in F^{-1}(y - g(z^{j-1}(y))) \,\middle|\, \|x - z^{j-1}(y)\| \le \alpha\nu\|z^{j-1}(y) - z^{j-2}(y)\| \right\}$$

for $y \in I\!B_\tau(\bar{y})$ and $M_j(y) = \emptyset$ for $y \notin I\!B_\tau(\bar{y})$, where we put $z^{-1}(y) = \bar{x}$ for $y \in I\!B_\tau(\bar{y})$. Then for $y \in I\!B_\tau(\bar{y})$ we obtain

$$\|z^j(y) - z^{j-1}(y)\| \le (\alpha\nu)^{j-1}\|z^1(y) - z^0(y)\| \le (\alpha\nu)^j\|z^0(y) - \bar{x}\|, \quad j = 2, \ldots, n.$$

Therefore,

$$\begin{aligned} \|z^j(y) - \bar{x}\| &\le \sum_{i=0}^{j} \|z^i(y) - z^{i-1}(y)\| \\ &\le \sum_{i=0}^{j} (\alpha\nu)^i\|z^0(y) - \bar{x}\| \le \frac{\lambda}{1 - \alpha\nu}\|y - \bar{y}\| \le \gamma\|y - \bar{y}\|. \end{aligned}$$

Hence, using (18), for $j = 1, 2, \ldots, n$ we have

$$\|z^j(y) - \bar{x}\| \le a \tag{21}$$

and also

$$\|y - g(z^j(y)) - \bar{y}\| \le \tau + \nu\|z^j(y) - \bar{x}\| \le \tau + \frac{\lambda\nu\tau}{1 - \alpha\nu} \le \frac{\tau}{1 - \alpha\nu} \le \beta \le b. \tag{22}$$

Consider the mapping $y \mapsto M_{n+1}(y)$, where

$$M_{n+1}(y) = \left\{ x \in F^{-1}(y - g(z^n(y))) \,\middle|\, \|x - z^n(y)\| \le \alpha\nu\|z^n(y) - z^{n-1}(y)\| \right\}$$

for $y \in I\!B_\tau(\bar{y})$ and $M_{n+1}(y) = \emptyset$ for $y \notin I\!B_\tau(\bar{y})$. As in the first step, we find that M_{n+1} is nonempty-closed-convex-valued. Let $y \in I\!B_\tau(\bar{y})$ and $x \in M_{n+1}(y)$, and let $y^k \in I\!B_\tau(\bar{y})$, $y^k \to y$ as $k \to \infty$. If $z^{n-1}(y) = z^n(y)$, then $M_{n+1}(y) = \{z^n(y)\}$, and consequently $x = z^n(y)$; then from

$$z^n(y^k) \in F^{-1}(y^k - g(z^{n-1}(y^k))) \cap \mathbb{B}_a(\bar{x}) \quad \text{and} \quad y^k - g(z^{n-1}(y^k)) \in \mathbb{B}_b(\bar{y}),$$

using the Aubin property of F^{-1}, we obtain that there exists $x^k \in F^{-1}(y^k - g(z^n(y^k)))$ such that

$$\|x^k - z^n(y^k)\| \le \lambda \|g(z^n(y^k)) - g(z^{n-1}(y^k))\| \le \alpha\nu \|z^n(y^k) - z^{n-1}(y^k)\|.$$

Therefore $x^k \in M_{n+1}(y^k)$, $x^k \to z^n(y) = x$ as $k \to \infty$, and hence M_{n+1} is inner semicontinuous in the case considered.

Let $z^n(y) \ne z^{n-1}(y)$. From (21) and (22) for $y = y^k$, since

$$z^n(y^k) \in F^{-1}(y^k - g(z^{n-1}(y^k))) \cap \mathbb{B}_a(\bar{x}),$$

the Aubin property of F^{-1} implies the existence of $\check{x}^k \in F^{-1}(y^k - g(z^n(y^k)))$ such that

$$\|\check{x}^k - z^n(y^k)\| \le \lambda \|g(z^n(y^k)) - g(z^{n-1}(y^k))\| \le \lambda\nu \|z^n(y^k) - z^{n-1}(y^k)\|.$$

Similarly, since $x \in F^{-1}(y - g(z^n(y))) \cap \mathbb{B}_a(\bar{x})$, there exists $\tilde{x}^k \in F^{-1}(y^k - g(z^n(y^k)))$ such that

$$\begin{aligned} \|\tilde{x}^k - x\| &\le \lambda(\|y^k - y\| + \|g(z^n(y^k)) - g(z^n(y))\|) \\ &\le \lambda(\|y^k - y\| + \nu\|z^n(y^k) - z^n(y)\|) \to 0 \text{ as } k \to \infty. \end{aligned}$$

Put

$$\varepsilon^k := \frac{\alpha\nu\|z^{n-1}(y) - z^{n-1}(y^k)\| + (1+\alpha\nu)\|z^n(y) - z^n(y^k)\| + \|\tilde{x}^k - x\|}{\alpha\nu\|z^n(y) - z^{n-1}(y)\| - \lambda\nu\|z^n(y^k) - z^{n-1}(y^k)\|}.$$

Then $\varepsilon^k \to 0$ as $k \to \infty$, hence $\varepsilon_k \in (0, 1)$ for large k. Taking

$$x^k = \varepsilon^k \check{x}^k + (1 - \varepsilon^k)\tilde{x}^k,$$

we obtain that $x^k \in F^{-1}(y^k - g(z^n(y^k)))$ for large k. Further, we can estimate $\|x^k - z^n(y^k)\|$ in the same way as in the first step, that is,

$$\begin{aligned}
\|x^k - z^n(y^k)\| &\le \varepsilon^k \|\check{x}^k - z^n(y^k)\| + (1-\varepsilon^k)\|\tilde{x}^k - z^n(y^k)\| \\
&\le \varepsilon^k \lambda\nu \|z^n(y^k) - z^{n-1}(y^k)\| \\
&\quad +(1-\varepsilon^k)(\|\tilde{x}^k - x\| + \|x - z^n(y)\| + \|z^n(y) - z^n(y^k)\|) \\
&\le \varepsilon^k \lambda\nu \|z^n(y^k) - z^{n-1}(y^k)\| + \|\tilde{x}^k - x\| \\
&\quad +(1-\varepsilon^k)\alpha\nu\|z^n(y) - z^{n-1}(y)\| + \|z^n(y) - z^n(y^k)\| \\
&\le \alpha\nu\|z^n(y^k) - z^{n-1}(y^k)\| + \alpha\nu\|z^n(y^k) - z^n(y)\| \\
&\quad +\alpha\nu\|z^{n-1}(y^k) - z^{n-1}(y)\| + \|\tilde{x}^k - x\| \\
&\quad +\|z^n(y) - z^n(y^k)\| - \varepsilon^k\alpha\nu\|z^n(y) - z^{n-1}(y)\| \\
&\quad +\varepsilon^k\lambda\nu\|z^n(y^k) - z^{n-1}(y^k)\| \\
&\le \alpha\nu\|z^n(y^k) - z^{n-1}(y^k)\| + \|\tilde{x}^k - x\| \\
&\quad +(1+\alpha\nu)\|z^n(y) - z^n(y^k)\| + \alpha\nu\|z^{n-1}(y) - z^{n-1}(y^k)\| \\
&\quad -\varepsilon^k(\alpha\nu\|z^n(y) - z^{n-1}(y)\| - \lambda\nu\|z^n(y^k) - z^{n-1}(y^k)\|) \\
&= \alpha\nu\|z^n(y^k) - z^{n-1}(y^k)\|.
\end{aligned}$$

We conclude that $x^k \in M_{n+1}(y^k)$, and since $x^k \to x$ as $k \to \infty$, the mapping M_{n+1} is inner semicontinuous on $\mathbb{B}_\tau(\bar{y})$. Hence, the mapping M_{n+1} has a continuous selection $z^{n+1} : \mathbb{B}_\tau(\bar{y}) \to X$. This selection satisfies

$$z^{n+1}(y) \in F^{-1}(y - g(z^n(y))) \text{ and } \|z^{n+1}(y) - z^n(y)\| \le \alpha\nu\|z^n(y) - z^{n-1}(y)\|,$$

therefore

$$\|z^{n+1}(y) - z^n(y)\| \le (\alpha\nu)^{(n+1)}\|z^0(y) - \bar{x}\|.$$

The induction step is now complete. In consequence, we have an infinite sequence of continuous functions $z^0, \ldots, z^n, \ldots$ such that for all $y \in \mathbb{B}_\tau(\bar{y})$ and for all n,

$$\|z^n(y) - \bar{x}\| \le \sum_{i=0}^{n}(\alpha\nu)^i\|z^0(y) - \bar{x}\| \le \frac{\lambda}{1-\alpha\nu}\|y - \bar{y}\| \le \gamma\|y - \bar{y}\|$$

and moreover,

$$\sup_{y \in \mathbb{B}_\tau(\bar{y})} \|z^{n+1}(y) - z^n(y)\| \le (\alpha\nu)^n \sup_{y \in \mathbb{B}_\tau(\bar{y})} \|z^0(y) - \bar{x}\| \le (\alpha\nu)^n\lambda\tau \quad \text{for } n \ge 1.$$

The sequence $\{z^n\}$ is a Cauchy sequence in the Banach space of functions that are continuous on $\mathbb{B}_\tau(\bar{y})$ equipped with the supremum norm. Then this sequence has a limit s which is a continuous function on $\mathbb{B}_\tau(\bar{y})$ and satisfies

$$s(y) \in F^{-1}(y - g(s(y)))$$

and

$$\|s(y) - \bar{x}\| \le \frac{\lambda}{1-\alpha\nu}\|y - \bar{y}\| \le \gamma\|y - \bar{y}\| \quad \text{for all } y \in \mathbb{B}_\tau(\bar{y}).$$

Thus, we obtain that s is a continuous local selection of $(g + F)^{-1}$ which has the calmness property (13). The proof is complete. □

Exercise 13.8. *Prove Theorem 13.6.*

Guide. Use an argument parallel to the proof of Theorem 13.5. First choose γ as required and then λ, α, and ν such that $\kappa < \lambda < \alpha < \nu^{-1}$ and $\nu > \mu$, and also

$$\frac{\lambda}{1-\alpha\nu} < \gamma.$$

There are neighborhoods U, V and Q of $\bar{x}$, 0 and $\bar{p}$, respectively, which are associated with the metric regularity of F at $\bar{x}$ for 0 with constant λ and the Lipschitz continuity of f with respect to x with constant ν uniformly in p. By appropriately choosing a sufficiently small radius τ of a ball around $\bar{p}$, construct an infinite sequence of continuous functions $z^k : I\!B_\tau(\bar{p}) \to X$, $k = 0, 1, \ldots$, which are uniformly convergent on $I\!B_\tau(\bar{p})$ to a function s satisfying the desired conditions. The initial z^0 satisfies

$$z^0(p) \in F^{-1}(-f(p, \bar{x})) \quad \text{and} \quad \|z^0(p) - \bar{x}\| \le \lambda\|f(p, \bar{x}) - f(\bar{p}, \bar{x})\|.$$

For $k = 1, 2, \ldots$, the function z^k is a continuous selection of the mapping

$$M_k : p \mapsto \left\{x \in F^{-1}(-f(p, z^{k-1}(p))) \,\middle|\, \|x - z^{k-1}(p)\| \le \alpha\nu\|z^{k-1}(p) - z^{k-2}(p)\|\right\}$$

for $p \in I\!B_\tau(\bar{p})$, where $z^{-1}(p) = \bar{x}$. Then for all $p \in I\!B_\tau(\bar{p})$ we obtain

$$z^k(p) \in F^{-1}(-f(p, z^{k-1}(p))) \text{ and } \|z^k(p) - z^{k-1}(p)\| \le (\alpha\nu)^k\|z^0(p) - \bar{x}\|,$$

hence,

$$\|z^k(y) - \bar{x}\| \le \frac{\lambda}{1-\alpha\nu}\|f(p, \bar{x}) - f(\bar{p}, \bar{x})\|.$$

The sequence $\{z^k\}$ is a Cauchy sequence of continuous functions, hence it is convergent with respect to the supremum norm. In the limit with $k \to \infty$, we obtain a selection s with the desired properties. □

Proof of Theorem 13.7. Let F and f satisfy the assumptions of the theorem. Choose γ to satisfy (16) and then pick $\kappa > \operatorname{reg}(F; \bar{x}\,|\,\bar{y})$ and $\nu > \widehat{\operatorname{clm}}_x(f; (\bar{p}, \bar{x}))$ such that $\kappa\nu < 1$ and $\gamma > \kappa(1 - \kappa\nu)$. As in the proof of Theorem 13.5, choose positive constants a and b such that

$$F^{-1}(y) \cap I\!B_a(\bar{x}) \text{ is a closed set,}$$

and

$$d(x, F^{-1}(y)) \le \kappa d(y, F(x)) \text{ for all } x \in I\!B_a(\bar{x}) \text{ and } y \in I\!B_b(\bar{y}).$$

Make a and b smaller if necessary so that $\min\{a, b\} < c$ and F^{-1} has the Aubin property at $\bar{y}$ for $\bar{x}$ with constants a, b, and κ; that is,

$$F^{-1}(y) \cap I\!B_a(\bar{x}) \subset F^{-1}(y') + \kappa\|y - y'\|I\!B \text{ for all } y, y' \in I\!B_b(\bar{y}),$$

and also

$$F^{-1}(y) \cap I\!B_a(\bar{x}) \neq \emptyset \quad \text{for all } y \in I\!B_b(\bar{y}).$$

Make a even smaller if necessary and choose a neighborhood Q of $\bar{p}$ such that

$$\|f(p,x) - f(p,\bar{x})\| \leq \nu\|x - \bar{x}\| \quad \text{for all } x \in I\!B_a(\bar{x}) \text{ and } p \in Q.$$

Then pick positive α and β such that

$$\alpha > \kappa, \quad \alpha\nu < 1, \quad \frac{\alpha}{1-\alpha\nu} < \gamma \text{ and } \beta < \min\left\{\frac{a}{\alpha}, b\right\}. \tag{23}$$

Choose a positive $\delta < a$ such that $\nu\delta < \beta$, and then a neighborhood Q of $\bar{p}$ such that

$$\|f(p,\bar{x}) - f(\bar{p},\bar{x})\| \leq \min\left\{\beta - \nu\delta, \frac{\delta(1-\alpha\nu)}{\alpha}\right\}.$$

For $x \in I\!B_\delta(\bar{x})$ and $p \in Q$ we have

$$\| - f(p,x) - \bar{y}\| \leq \|f(p,x) - f(\bar{p},\bar{x})\| \leq \nu\delta + \|f(p,\bar{x}) - f(\bar{p},\bar{x})\| \leq \beta.$$

Fix $p \in Q$ and consider the function

$$I\!B_\delta(\bar{x}) \ni x \mapsto \Phi_p(x) = s(-f(p,x)),$$

where s is the continuous selection whose existence is established in Proposition 13.4. Since $-f(\bar{p},\bar{x}) = \bar{y} \in F(\bar{x})$, we have

$$\begin{aligned}
\|\Phi_p(x) - \bar{x}\| &= \|s(-f(p,x)) - \bar{x}\| \leq \alpha d(-f(p,x), F(\bar{x})) \\
&\leq \alpha\| - f(p,x) - (-f(\bar{p},\bar{x}))\| \\
&\leq \alpha(\|f(p,x) - f(p,\bar{x})\| + \alpha\|f(p,\bar{x}) - f(\bar{p},\bar{x})\| \\
&\leq \alpha\nu\delta + \alpha\frac{\delta(1-\alpha\nu)}{\alpha} = \delta.
\end{aligned}$$

Hence, Φ_p maps $I\!B_\delta(\bar{x})$ into itself. Therefore, by the Brouwer fixed point theorem, there exists a fixed point $\sigma(p) = \Phi_p(\sigma(p)) \in I\!B_\delta(\bar{x})$. But then

$$\sigma(p) = s(-f(p,\sigma(p))) \in F^{-1}(-f(p,\sigma(p))) \cap I\!B_\delta(\bar{x}),$$

that is,

$$0 \in f(p,\sigma(p)) + F(\sigma(p)),$$

and also

$$\|\sigma(p) - \bar{x}\| = \alpha d(-f(p,\sigma(p)), F(\bar{x})).$$

Note that

$$\begin{aligned}
d(-f(p,\sigma(p)), F(\bar{x})) &\leq \|f(p,\sigma(p)) - f(p,\bar{x})\| + d(-f(p,\bar{x}), F(\bar{x})) \\
&\leq \nu\|\sigma(p) - \bar{x}\| + d(-f(p,\bar{x}), F(\bar{x})).
\end{aligned}$$

Hence,

$$d(-f(p, \sigma(p)), F(\bar{x})) \leq \frac{\alpha}{1-\alpha\nu} d(-f(p, \bar{x}), F(\bar{x})),$$

which gives us (17) in view of the choice of the constants ν and α in (23). This completes the proof. □

Lecture 14
Radius of Regularity

A classical result about matrices, often referred to as the Eckart–Young theorem, says that for any nonsingular matrix A,

$$\inf\left\{ \|B\| \,\middle|\, A + B \text{ is singular} \right\} = \frac{1}{\|A^{-1}\|}. \tag{1}$$

In other words, the distance from a nonsingular matrix A to the set of singular matrices equals $\|A^{-1}\|^{-1}$; that is, the quantity $\|A^{-1}\|^{-1}$ gives the *radius of nonsingularity* around A (Fig. 14.1). More broadly, this result can be regarded as a *radius theorem*, furnishing a bound on how far perturbations of some sort in the specification of a problem can go before some key property is lost. The radius of nonsingularity is tied with the *condition number* of a matrix, but we will not go into that here. Radius theorems can be of course considered not only for equations, linear and nonlinear, but also generalized equations, optimization problems, etc. We start down that track by stating and proving a more general version of the equality (1) for bounded linear mappings acting in Banach spaces.

Theorem 14.1. *Let X and Y be Banach spaces and let $A \in \mathcal{L}(X, Y)$ be invertible. Then*

$$\inf_{B \in \mathcal{L}(X,Y)} \left\{ \|B\| \,\middle|\, A + B \text{ is not invertible} \right\} = \frac{1}{\|A^{-1}\|}. \tag{2}$$

Moreover the infimum is the same if restricted to mappings B of rank one.

Proof. We will show first that for any $B \in \mathcal{L}(X, Y)$ with $\|A^{-1}\|{\cdot}\|B\| < 1$ one has

$$\|(A + B)^{-1}\| \leq \frac{\|A^{-1}\|}{1 - \|A^{-1}\|\|B\|}. \tag{3}$$

© The Author(s), under exclusive license to Springer Nature Switzerland AG 2021

A. L. Dontchev, *Lectures on Variational Analysis*, Applied Mathematical Sciences 205, https://doi.org/10.1007/978-3-030-79911-3_14

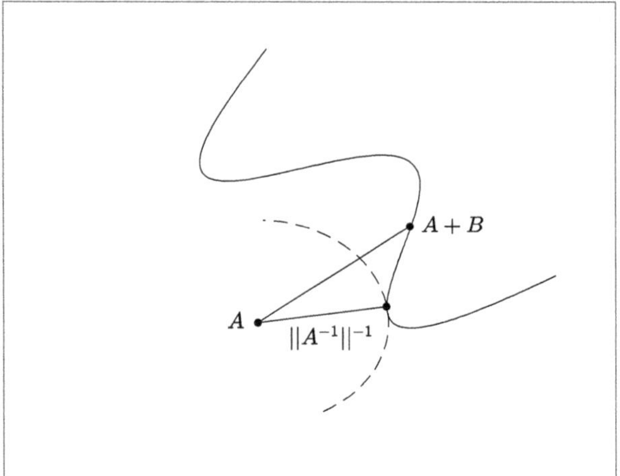

Fig. 14.1: Radius of regularity

Let $C = BA^{-1}$; then $||C|| < 1$ and hence $||C^n|| \leq ||C||^n \longrightarrow 0$ as $n \longrightarrow \infty$. Also, the mappings

$$S_n = \sum_{i=0}^{n} C^i \quad \text{for } n = 0, 1, \ldots$$

form a Cauchy sequence in the Banach space $\mathcal{L}(X, Y)$ which therefore converges to some $S \in \mathcal{L}(X, Y)$. Observe that, for each n,

$$S_n(I - C) = I - C^{n+1} = (I - C)S_n,$$

and hence, through passing to the limit, one has $S = (I - C)^{-1}$. On the other hand

$$||S_n|| \leq \sum_{i=0}^{n} ||C^i|| \leq \sum_{i=0}^{\infty} ||C||^i = \frac{1}{1 - ||C||}.$$

Thus, we obtain

$$||(I - C)^{-1}|| \leq \frac{1}{1 - ||C||}.$$

All that remains is to bring in the identity $(I - C)A = A - B$ and the inequality $||C|| \leq ||A^{-1}|| ||B||$, and to observe that the sign of B does not matter. This gives us (3) which implies "$\geq$" in (2).

To obtain the opposite inequality and thereby complete the proof, we take any $r > 1/||A^{-1}||$ and construct a mapping B of rank one such that $A + B$ is not invertible and $||B|| < r$. There exists $\hat{x}$ with $||A\hat{x}|| = 1$ and $||\hat{x}|| > 1/r$. Choose an $x^* \in X^*$ such that $x^*(\hat{x}) = ||\hat{x}||$ and $||x^*|| = 1$. The linear and bounded mapping

$$Bx = -\frac{x^*(x)A\hat{x}}{\|\hat{x}\|} \tag{4}$$

has $\|B\| = 1/\|\hat{x}\|$ and $(A + B)\hat{x} = A\hat{x} - A\hat{x} = 0$. Then $A + B$ is not invertible and hence the infimum in (3) is $\leq r$. It remains to note that B in (4) is of rank one. □

We will now extend Theorem 14.1 for positively homogeneous mappings $H : X \rightrightarrows Y$; here and further on, X and Y are Banach spaces. Recall that a mapping $H : X \rightrightarrows Y$ is said to be positively homogeneous when its graph is a cone in $X \times Y$. In the preparatory section, and then again in Lecture 7, we defined the outer norm as follows:

$$\|H\|^+ = \sup_{\|x\|\leq 1} \sup_{y\in H(x)} \|y\|.$$

When $\operatorname{dom} H = X$ and H is single-valued, this expression reduces to the usual operator norm $\|H\|$.

Proposition 14.2. *The outer norm of a positively homogeneous mapping $H : X \rightrightarrows Y$ satisfies*

$$\|H\|^+ = \inf\left\{ \kappa \in (0,\infty) \,\middle|\, H(\mathbb{B}) \subset \kappa\mathbb{B} \right\} = \sup_{\|y\|=1} \frac{1}{d(0, H^{-1}(y))}, \tag{5}$$

and moreover

$$\|H\|^+ < \infty \implies H(0) = \{0\}, \tag{6}$$

with this implication becoming an equivalence when H has closed graph and $\dim X < \infty$.

Recall that for a positively homogeneous $H : X \rightrightarrows Y$ and a linear $B : X \to Y$, we have $(H + B)(x) = H(x) + Bx$ for every $x \in X$. It will be convenient to introduce the following definition:

Nonsingularity of Positively Homogeneous Mappings. *A positively homogeneous mapping $H : X \rightrightarrows Y$ is said to be nonsingular if $\|H^{-1}\|^+ < \infty$; it is said to be singular if $\|H^{-1}\|^+ = \infty$.*

From Proposition 14.2, nonsingularity of H in this sense implies that $H^{-1}(0) = \{0\}$; moreover, when $\dim X < \infty$ and $\operatorname{gph} H$ is closed the converse is true as well.

We are now ready to state the following radius theorem for nonsingularity of positively homogeneous mappings:

Theorem 14.3. *For any $H : X \rightrightarrows Y$ that is positively homogeneous and nonsingular, one has*

$$\inf_{B\in\mathcal{L}(X,Y)} \left\{ \|B\| \,\middle|\, H + B \text{ is singular} \right\} = \frac{1}{\|H^{-1}\|^+}. \tag{7}$$

Moreover the infimum is the same if restricted to mappings B of rank one.

We will show next that, in finite dimensions at least, the radius result in Theorem 14.3 is valid when H is replaced by *any set-valued mapping* F whose graph is locally closed around the reference pair $(\bar{x}, \bar{y})$.

Theorem 14.4. *Let X and Y be linear normed spaces, and for $F : X \rightrightarrows Y$ and $\bar{y} \in F(\bar{x})$ let* gph F *be metrically regular at $\bar{x}$ for $\bar{y}$. Then*

$$\inf_{B\in\mathcal{L}(X,Y)} \left\{ \|B\| \,\middle|\, F + B \text{ is not metrically regular at } \bar{x} \text{ for } \bar{y} + B\bar{x} \right\} \geq \frac{1}{\operatorname{reg}(F; \bar{x}\,|\,\bar{y})}. \tag{8}$$

In fact, when X and Y are finite-dimensional, we have equality in (8)*. Moreover, the infimum in the left side of* (8) *is unchanged if taken with respect to linear mappings of rank 1, but also remains unchanged when the class of perturbations B is enlarged to all locally Lipschitz continuous functions g, with $\|B\|$ replaced by the Lipschitz modulus* $\operatorname{lip}(g; \bar{x})$ *of g at $\bar{x}$.*

Proof. The general perturbation inequality derived in Theorem 6.2 produces the estimate

$$\inf_{g:X\to Y} \left\{ \operatorname{lip}(g; \bar{x}) \,\middle|\, F + g \text{ is not metrically regular at } \bar{x} \text{ for } \bar{y} + g(\bar{x}) \right\} \geq \frac{1}{\operatorname{reg}(F; \bar{x}\,|\,\bar{y})}. \tag{9}$$

Indeed, the perturbation B in our case has Lipschitz constant $\mu = \|B\|$. Note that (9) becomes the equality (8) in the case when $\operatorname{reg}(F; \bar{x}\,|\,\bar{y}) = 0$ under the convention $1/0 = \infty$. To confirm the opposite inequality when X and Y are finite-dimensional and $\operatorname{reg}(F; \bar{x}\,|\,\bar{y}) > 0$, we apply the coderivative criterion for metric regularity given in Theorem 8.5, according to which F is metrically regular if and only if $\|D^*F(\bar{x}\,|\,\bar{y})\|^+$ is finite which, according to Proposition 14.2, is equivalent to nonsingularity of the coderivative mapping $D^*F(\bar{x}\,|\,\bar{y})$. Thus, all we need to prove is that

$$\inf_{B\in\mathcal{L}(X,Y)} \left\{ \|B\| \,\middle|\, D^*(F + B)(\bar{x}\,|\,\bar{y} + B\bar{x}) \text{ is singular} \right\} \leq \frac{1}{\|D^*F(\bar{x}\,|\,\bar{y})^{-1}\|^+}. \tag{10}$$

By the definition of the coderivative, it is elementary that

$$D^*(F + B)(\bar{x}\,|\,\bar{y} + B\bar{x}) = D^*F(\bar{x}\,|\,\bar{y}) + B^*.$$

Also, B^* is represented by the transposed matrix B^{T} and $\|B\| = \|B^{\mathsf{T}}\|$. Thus (10) comes down to

$$\inf_{B\in\mathcal{L}(X,Y)} \left\{ \|B\| \,\middle|\, D^*F(\bar{x}\,|\,\bar{y}) + B^* \text{ is singular} \right\} \leq \frac{1}{\|D^*F(\bar{x}\,|\,\bar{y})^{-1}\|^+}. \tag{11}$$

Furthermore, again by the finite dimensionality, every mapping in $\mathcal{L}(Y^*, X^*)$ is the adjoint B^* of some $B \in \mathcal{L}(X, Y)$. Therefore (11) follows from Theorem 14.3 for $H = D^*F(\bar{x}\,|\,\bar{y})$. □

It is now easy to obtain a parallel radius theorem for strong regularity.

Theorem 14.5. *For finite-dimensional normed linear spaces X and Y, let $F : X \rightrightarrows Y$ have $\bar{y} \in F(\bar{x})$. Suppose that F is strongly regular at $\bar{x}$ for $\bar{y}$. Then*

$$\inf_{B\in\mathcal{L}(X,Y)} \left\{ \|B\| \;\middle|\; F + B \textit{ is not strongly regular at } \bar{x} \textit{ for } \bar{y} + B\bar{x} \right\} = \frac{1}{\operatorname{reg}(F;\bar{x}\,|\,\bar{y})}. \tag{12}$$

Moreover, the infimum is unchanged if taken with respect to linear mappings of rank 1, but also remains unchanged when the class of perturbations B is enlarged to the class of locally Lipschitz continuous functions g with $\|B\|$ replaced by the Lipschitz modulus $\operatorname{lip}(g;\bar{x})$.

Proof. Theorem 8.6 implies that "$\geq$" holds in (12) when the linear perturbation is replaced by a Lipschitz perturbation, and moreover that (12) is satisfied in the limit case $\operatorname{reg}(F;\bar{x}\,|\,\bar{y}) = 0$ under the convention $1/0 = \infty$. The inequality becomes an equality with the observation that the assumed strong metric regularity of F implies that F has locally closed graph at $(\bar{x},\bar{y})$ and is metrically regular at $\bar{x}$ for $\bar{y}$. Hence the infimum in (12) is not greater than the infimum in (8). □

Next comes a radius theorem for strong subregularity to go along with the ones for metric regularity and strong regularity.

Theorem 14.6. *Let X and Y be finite-dimensional normed linear spaces and for $F : X \rightrightarrows Y$ and $\bar{y} \in F(\bar{x})$ let* gph F *be locally closed at $(\bar{x},\bar{y})$. Suppose that F is strongly subregular at $\bar{x}$ for $\bar{y}$. Then*

$$\inf_{B\in\mathcal{L}(X,Y)} \left\{ \|B\| \;\middle|\; F + B \textit{ is not strongly subregular at } \bar{x} \textit{ for } \bar{y} + B\bar{x} \right\} = \frac{1}{\operatorname{subreg}(F;\bar{x}\,|\,\bar{y})}.$$

Moreover, the infimum is unchanged if taken with respect to mappings B of rank 1, but also remains unchanged when the class of perturbations is enlarged to the class of functions $g : X \to Y$ that are calm at $\bar{x}$ and continuous around $\bar{x}$, with $\|B\|$ replaced by the calmness modulus $\operatorname{clm}(g;\bar{x})$.

Proof. From the equivalence of the strong subregularity of a mapping F at $\bar{x}$ for $\bar{y}$ with the nonsingularity of its graphical derivative $DF(\bar{x}\,|\,\bar{y})$, as shown in Theorem 12.7, we have

$$\begin{aligned}\inf_{B\in\mathcal{L}(X,Y)} &\left\{ \|B\| \;\middle|\; F + B \text{ is not strongly subregular at } \bar{x} \text{ for } \bar{y} + B\bar{x} \right\} \\ &= \inf_{B\in\mathcal{L}(X,Y)} \left\{ \|B\| \;\middle|\; D(F+B)(\bar{x}\,|\,\bar{y}+B\bar{x}) \text{ is singular} \right\}.\end{aligned}$$

To get the desired equality, it is sufficient to repeat the proof of Theorem 14.4 with the coderivative replaced by the graphical derivative. □

Proofs

Proof of Proposition 14.2. We obtain the first part of (5) simply by rewriting the formula in terms of the unit ball and utilizing the positive homogeneity. The infimum in the middle part of (5) is unchanged when y is restricted to have $\|y\| = 1$, and in this way it can be identified with the infimum of all $\kappa \in (0, \infty)$ such that $\kappa \geq 1/\|x\|$ whenever $x \in H^{-1}(y)$ and $\|y\| = 1$. (It is correct in this to interpret $1/\|x\| = \infty$ when $x = 0$.) This shows that the middle expression in (5) agrees with the final one.

Moving on to (6), we observe that when $\|H\|^+ < \infty$ the middle expression in (5) implies that if $(0, y) \in \text{gph } H$ then y must be 0. To prove the converse implication we will need the assumption that gph H is closed and the unit ball is compact, which is the same as $\dim X < \infty$. Suppose that $H(0) = \{0\}$. If $\|H\|^+ = \infty$, there has to be a sequence of points $(x_k, y_k) \in \text{gph } H$ such that $0 < \|y_k\| \to \infty$ but x_k is bounded. Consider then the sequence of pairs (w_k, u_k) in which $w_k = x_k/\|y_k\|$ and $u_k = y_k/\|y_k\|$. We have that $w_k \to 0$, while u_k has a cluster point $\bar{u}$ with $\|\bar{u}\| = 1$. Moreover, $(w_k, u_k) \in \text{gph } H$ by the positive homogeneity and hence, through the closedness of gph H, we must have $H(0) \ni \bar{u}$. This contradicts our assumption that $H(0) = \{0\}$ and completes the proof of (6). □

Proof of Theorem 14.3. We first show that for a positively homogeneous mapping $H : X \rightrightarrows Y$ be positively with $\|H^{-1}\|^+ < \infty$ and any $B \in \mathcal{L}(X, Y)$ with the property that $\|H^{-1}\|^+\|B\| < 1$, one has

$$\|(H + B)^{-1}\|^+ \leq \frac{\|H^{-1}\|^+}{1 - \|H^{-1}\|^+\|B\|}. \tag{13}$$

Having $\|H^{-1}\|^+ = 0$ is equivalent to having $\text{dom } H = \{0\}$; in this case, since $\emptyset + y = \emptyset$ for any y, we get $\|(H+B)^{-1}\|^+ = 0$ for any $B \in \mathcal{L}(X, Y)$ as claimed. Suppose therefore instead that $0 < \|H^{-1}\|^+ < \infty$. If the estimate (13) is false, there is some $B \in \mathcal{L}(X, Y)$ with $\|B\| < [\|H^{-1}\|^+]^{-1}$ such that $\|(H+B)^{-1}\|^+ > ([\|H^{-1}\|^+]^{-1} - \|B\|)^{-1}$. In particular $B \neq 0$ then, and by definition there must exist $y \in I\!B$ and $x \in (H + B)^{-1}(y)$ such that $\|x\| > ([\|H^{-1}\|^+]^{-1} - \|B\|)^{-1}$, which is the same as

$$\frac{1}{\|x\|^{-1} + \|B\|} > \|H^{-1}\|^+. \tag{14}$$

But then $y - Bx \in H(x)$ and

$$\|y - Bx\| \leq \|y\| + \|B\|\|x\| \leq 1 + \|B\|\|x\|. \tag{15}$$

If $y = Bx$ then $0 \in H(x)$, so (6) yields $x = 0$, a contradiction. Hence $\alpha := \|y - Bx\|^{-1} > 0$, and due to the positive homogeneity of H we have $(\alpha x, \alpha(y - Bx)) \in \text{gph } H$ and $\alpha\|y - Bx\| = 1$, which implies, by definition,

$$\|H^{-1}\|^+ \geq \frac{\|x\|}{\|y - Bx\|}.$$

Combining this inequality with (14) and (15), we get

$$\|H^{-1}\|^{+} \geq \frac{\|x\|}{\|y - Bx\|} \geq \frac{\|x\|}{1 + \|B\|\|x\|} = \frac{1}{\|x\|^{-1} + \|B\|} > \|H^{-1}\|^{+}.$$

This is impossible, and hence (13) is satisfied.

Continuing with the proof of (7), from (13) we get "≥" in (7), and also "=" for the case $\|H^{-1}\|^{+} = 0$ under the convention $1/0 = \infty$. Let $\|H^{-1}\|^{+} > 0$ and consider any $r > 1/\|H^{-1}\|^{+}$. There exists $(\hat{x}, \hat{y}) \in \text{gph } H$ with $\|\hat{y}\| = 1$ and $\|\hat{x}\| > 1/r$. Let $x^* \in X^*$, $x^*(\hat{x}) = \|\hat{x}\|$ and $\|x^*\| = 1$. The linear and bounded mapping

$$Bx = -\frac{x^*(x)\hat{y}}{\|\hat{x}\|}$$

has $\|B\| = 1/\|\hat{x}\| < r$ and $(H + B)(\hat{x}) = H(\hat{x}) - \hat{y} \ni 0$. Then the nonzero vector $\hat{x}$ belongs to $(H+B)^{-1}(0)$, hence $\|(H+B)^{-1}\|^{+} = \infty$, i.e., $H+B$ is singular. The infimum in (3) must therefore be less than r. Appealing to the choice of r, we conclude that the infimum in (3) cannot be more than $1/\|H^{-1}\|^{+}$, and we are done. □

Lecture 15
Newton Method for Generalized Equations

In this lecture, we consider a numerical method which goes back to Isaac Newton for solving the generalized equation

$$\text{find } x \text{ such that } f(x) + F(x) \ni 0, \tag{1}$$

where $f : X \to Y$ is a Fréchet differentiable function with derivative Df and $F : X \rightrightarrows Y$ is a set-valued mapping; here X and Y are Banach spaces. In the particular case of (1) when we have an equation, $f(x) = 0$, the classical Newton method generates a sequence of points x_k obtained by the iteration

$$f(x_k) + Df(x_k)(x_{k+1} - x_k) = 0, \quad k = 1, 2, \ldots \tag{2}$$

staring from a suitably chosen x_0. In numerical analysis textbooks, one can usually find a result saying that when the derivative $Df(\bar{x})$ is invertible at a solution $\bar{x}$ of the equation and the starting point x_0 is close to $\bar{x}$, then the Newton iteration (2) produces a sequence which converges to $\bar{x}$ in a certain way.

For generalized equations of the form (1), we consider the following version of the Newton method:

$$f(x_k) + Df(x_k)(x_{k+1} - x_k) + F(x_{k+1}) \ni 0, \quad \text{for } k = 0, 1, \ldots. \tag{3}$$

This method uses a linearization of f at the current point x_k. It reduces to the classical Newton method (2) for the case of a nonlinear equation, when F is the zero mapping. When (1) represents the optimality systems for a nonlinear programming problem, the iteration (3) becomes the popular sequential quadratic programming (SQP) algorithm, which is discussed later in this lecture.

Interestingly enough, Newton used this method for solving a third degree polynomial equation and attributed the method to F. Viéte. J. Raphson extended the method to polynomial equations of arbitrary degree and named it after Newton. It was T. Simpson who stated several decades later the method for an equation using Newton's fluxions as derivatives.

© The Author(s), under exclusive license to Springer Nature Switzerland AG 2021

A. L. Dontchev, *Lectures on Variational Analysis*, Applied Mathematical Sciences 205, https://doi.org/10.1007/978-3-030-79911-3_15

We will not discuss stopping criteria here; we assume that Newton method (3) generates an infinite sequence $\{x_k\}$ and if this sequence is convergent to a solution $\bar{x}$ of (1), then we say that the method is convergent. Note that this also covers the finite convergence: If $x_{k+1} = x_k$ for some k, then x_k is a solution and in this case we may extend the sequence to infinity by simply taking $x_{k+1} = x_k$ for all k.

We introduce next two modes of convergence. A sequence $\{x_k\}$ is said to be *superlinearly* convergent to $\bar{x}$ when there exist a natural N and sequence of reals $\{\alpha_k\}$ such that $\alpha_k \searrow 0$ and

$$\|x_{k+1} - \bar{x}\| \le \alpha_k \|x_k - \bar{x}\| \quad \text{for } k = N, N+1, \ldots.$$

A sequence $\{x_k\}$ is *quadratically* convergent to $\bar{x}$ when there exist a natural N and a positive constant c such that

$$\|x_{k+1} - \bar{x}\| \le c\|x_k - \bar{x}\|^2 \quad \text{for } k = N, N+1, \ldots.$$

In this lecture, we assume the following: The generalized equation (1) has a solution $\bar{x}$, the mapping F has closed graph, the function f is continuously differentiable in a neighborhood O of $\bar{x}$, and the derivative Df is Lipschitz continuous on O. Our first result shows that under metric regularity of the mapping $f + F$ at $\bar{x}$ for 0 the iteration (3) is sure to generate a sequence which is quadratically convergent to $\bar{x}$, provided that the starting point x_0 of the iteration (3) is sufficiently close to $\bar{x}$. Its proof, as well as the proofs of the other theorems in this lecture, are given at the end of lecture.

Theorem 15.1. *Consider the generalized equation* (1) *under the above assumptions and assume that the mapping $f + F$ is metrically regular at $\bar{x}$ for* 0*. Then there exists a neighborhood O of $\bar{x}$ such that for every starting point $x_0 \in O$ there exists a sequence $\{x_k\}$ generated by the iteration* (3) *whose elements are in O which is quadratically convergent to $\bar{x}$.*

The following theorems present convergence results that are parallel to Theorem 15.1 for the cases when $f + F$ is strongly regular and strongly subregular at $\bar{x}$ for 0.

Theorem 15.2. *Consider the generalized equation* (1) *under the standing assumptions and assume that the mapping $f + F$ is strongly subregular at $\bar{x}$ for* 0*. Then there exists a neighborhood O of $\bar{x}$ such that for every starting point $x_0 \in O$ every sequence generated by the iteration* (3) *which is convergent to $\bar{x}$ is in fact quadratically convergent to $\bar{x}$.*

In addition, if the mapping $f + F$ is not only strongly subregular but strongly regular at $\bar{x}$ for 0*, then there exists a neighborhood O of $\bar{x}$ such that for every starting point $x_0 \in O$ there exists a unique sequence $\{x_k\}$ generated by the iteration* (3) *whose elements are in O which is quadratically convergent to $\bar{x}$.*

As an application, we employ the Newton method to the nonlinear programming problem

$$\text{minimize } g_0(x) \text{ over all } x \text{ satisfying } g_i(x) \begin{cases} \leq 0 & \text{for } i \in [1, s], \\ = 0 & \text{for } i \in [s+1, m], \end{cases} \tag{4}$$

where the functions $g_i : I\!R^n \to I\!R$ are twice continuously differentiable. In terms of the Lagrange function

$$L(x, y) = g_0(x) + y_1 g_1(x) + \cdots + y_m g_m(x),$$

under the Mangasarian–Fromovitz constraint qualification, the first-order optimality (Karush–Kuhn–Tucker) condition takes the form

$$\begin{cases} \nabla_x L(x, y) = 0, \\ g(x) \in N_{I\!R^s_+ \times I\!R^{m-s}}(y), \end{cases} \tag{5}$$

where we denote by $g(x)$ the vector with components $g_1(x), \ldots, g_m(x)$. Let $\bar{x}$ be a local minimum for (4) and let $\bar{y}$ be an associated Lagrange multiplier vector. As applied to the variational inequality (5), the Newton method (3) consists in generating a sequence $\{(x_k, y_k)\}$ starting from a point (x_0, y_0), close enough to $(\bar{x}, \bar{y})$, according to the iteration

$$\begin{cases} \nabla_x L(x_k, y_k) + \nabla^2_{xx} L(x_k, y_k)(x_{k+1} - x_k) + \nabla g(x_k)^{\mathsf{T}}(y_{k+1} - y_k) = 0, \\ g(x_k) + \nabla g(x_k)(x_{k+1} - x_k) \in N_{I\!R^s_+ \times I\!R^{m-s}}(y_{k+1}). \end{cases} \tag{6}$$

Theorem 11.5 (with suppressed dependence on the parameter p in its statement) provides conditions under which the mapping appearing in the variational inequality (5) is strongly regular at the reference point: linear independence of the gradients of the active constraints and the strong second-order sufficient optimality condition. Under these conditions, we can find (x_{k+1}, y_{k+1}) which satisfies (6) by solving the quadratic programming problem

$$\begin{aligned} &\text{minimize } \Big[\frac{1}{2} \langle x - x_k, \nabla^2_{xx} L(x_k, y_k)(x - x_k) \rangle \\ &\qquad\qquad + \langle \nabla_x L(x_k, y_k) - \nabla g(x_k)^{\mathsf{T}} y_k, (x - x_k) \rangle \Big] \\ &\text{subject to } \; g_i(x_k) + \nabla g_i(x_k)(x - x_k) \begin{cases} \leq 0 & \text{for } i \in [1, s], \\ = 0 & \text{for } i \in [s+1, m]. \end{cases} \end{aligned} \tag{7}$$

Thus, under strong regularity of the mapping in (5), the Newton method (3) comes down to sequentially solving quadratic programs of the form (7). This specific application of the Newton method is therefore called the *sequential quadratic programming* (SQP) method.

We consider next a modification of the Newton method for solving the generalized equation (1) in finite dimensions, that is, for $X = Y = I\!R^n$, on the assumption that the function f is Lipschitz continuous around a solution $\bar{x}$ but not necessarily differentiable there. Then, for a Newton-type iteration for (1), instead of the derivative of f we use an element of Clarke's generalized Jacobian $\bar{\partial} f(\bar{x})$ introduced in Lecture 9. We will prove convergence of the method for a particular class of functions f called *semismooth*.

Semismooth Functions. *A function $f : \mathbb{R}^n \to \mathbb{R}^m$ is said to be semismooth at $\bar{x} \in \operatorname{int} \operatorname{dom} f$ when it is Lipschitz continuous around $\bar{x}$, directionally differentiable in every direction, and for every $\varepsilon > 0$ there exists $\delta > 0$ such that*

$$\|f(x) - f(\bar{x}) - A(x - \bar{x})\| \le \varepsilon \|x - \bar{x}\| \text{ for every } x \in \mathbb{B}_\delta(\bar{x}) \text{ and every } A \in \bar{\partial} f(x).$$

Let us recall the definition of the directional derivative given in the preparatory lecture. For $f : \mathbb{R}^n \to \mathbb{R}^m$, a point $\bar{x} \in \operatorname{dom} f$ and a vector $d \in \mathbb{R}^n$, the limit

$$f'(\bar{x}; d) = \lim_{t \searrow 0} \frac{f(\bar{x} + td) - f(\bar{x})}{t},$$

when it exists, is called the directional derivative of f at $\bar{x}$ for d. If this directional derivative exists for every $d \in \mathbb{R}^n$, f is said to be *directionally differentiable* at $\bar{x}$.

Examples.

(1) The function $f(x) = \|x\|$ is semismooth at 0 and smooth everywhere else.

(2) The function $f : \mathbb{R}^2 \to \mathbb{R}$ defined as $f(x) = x_1 + x_2 - \|x\|$ is semismooth at zero. This function is known as the Fischer–Burmeister function and is used to numerically handle complementarity problems.

(3) Every piecewise smooth function is semismooth in the interior of its domain.

(4) A Lipschitz continuous function $f : \mathbb{R}^n \to \mathbb{R}^m$ is semismooth at $\bar{x}$ if and only if for any $A \in \bar{\partial} f(\bar{x} + h)$ one has $Ah - f'(\bar{x}; h) = o(\|h\|)$.

We consider the following nonsmooth version of the Newton method for solving the generalized equation (1) with a semismooth function f:

$$f(x_k) + A_k(x_{k+1} - x_k) + F(x_{k+1}) \ni 0 \quad \text{for } A_k \in \bar{\partial} f(x_k), \quad k = 0, 1, \dots. \tag{8}$$

This method is usually referred to as the *semismooth Newton method*. The theorem given next shows superlinear convergence of this method.

Theorem 15.3. *Consider the iteration* (8) *applied to the generalized equation* (1) *with a solution $\bar{x}$ for a function f which is semismooth at $\bar{x}$. Assume that for every $A \in \bar{\partial} f(\bar{x})$ the mapping*

$$G_A : x \mapsto f(\bar{x}) + A(x - \bar{x}) + F(x)$$

is strongly regular at $\bar{x}$ for 0*. Then there exists a neighborhood O of $\bar{x}$ such that for every $x_0 \in O$ and $k = 0, 1, \dots$ there is a unique O iterate x_{k+1} obtained by the iteration* (8) *for any choice of $A_k \in \bar{\partial} f(x_k)$. Furthermore, the sequence $\{x_k\}$ is superlinearly convergent to $\bar{x}$.*

There are a number of delicate points in the above statements and many important questions arise, but we stop here.

Proofs

Proof of Theorem 15.1. There are a number of ways to prove this result; here we choose to present a proof based on the contraction mapping principle for set-valued mappings stated in Theorem 5.4.

According to Corollary 5.3, metric regularity of $f + F$ at $\bar{x}$ for 0 is equivalent to metric regularity of the mapping

$$x \mapsto G(x) = f(\bar{x}) + Df(\bar{x})(x - \bar{x}) + F(x). \tag{9}$$

Furthermore, we also know that metric regularity of G at $\bar{x}$ for 0 is equivalent to the Aubin property of G^{-1} at 0 for $\bar{x}$; that is, the existence of positive constants μ, a, and b such that

$$e(G^{-1}(y') \cap I\!B_a(\bar{x}), G^{-1}(y'')) \le \mu\|y' - y''\| \quad \text{for all } y', y'' \in I\!B_b(0). \tag{10}$$

Make a smaller if necessary such that $I\!B_a(\bar{x}) \subset O$, let L be a Lipschitz constant of Df over $I\!B_a(\bar{x})$, and choose $c > \frac{1}{2}\mu L$. Then pick $\delta > 0$ such that

$$\delta \le \min\left\{a, \sqrt{\frac{2b}{3L}}, \frac{1}{c}, \frac{1}{\mu L} - \frac{1}{2c}\right\}. \tag{11}$$

Choose any $x_0 \in I\!B_\delta(\bar{x})$. If $x_0 = \bar{x}$ there is nothing to prove; let $x_0 \neq \bar{x}$ and consider the mapping

$$x \mapsto \Phi_0(x) = G^{-1}(f(\bar{x}) + Df(\bar{x})(x - \bar{x}) - f(x_0) - Df(x_0)(x - x_0)).$$

In further lines we use the following standard estimate from calculus: if $x, u \in O$ then

$$\begin{aligned} &\|f(x) + Df(x)(u - x) - f(u)\| \\ &\quad = \left\|\int_0^1 Df(u + t(x - u))(x - u)dt - Df(x)(x - u)\right\| \\ &\quad \le \tfrac{L}{2}\|x - u\|^2. \end{aligned} \tag{12}$$

For any $x \in I\!B_\delta(\bar{x})$ we have

$$\begin{aligned} &\|f(\bar{x}) + Df(\bar{x})(x - \bar{x}) - f(x_0) - Df(x_0)(x - x_0)\| \\ &\quad \le \|f(\bar{x}) - f(x_0) - Df(x_0)(\bar{x} - x_0)\| \\ &\quad\quad + \|(Df(\bar{x}) - Df(\bar{x}))(x - \bar{x})\| \le \tfrac{3}{2}L\delta^2 < b, \end{aligned} \tag{13}$$

where in the last inequality we use (11). Utilizing the relations (10) through (13) we obtain

$$\begin{aligned}
d(\bar{x}, \Phi_0(\bar{x})) &\le e(G^{-1}(0) \cap I\!B_\delta(\bar{x}), G^{-1}(f(\bar{x}) - f(x_0) - Df(x_0)(\bar{x} - x_0)) \\
&\le \mu\|f(\bar{x}) - f(x_0) - Df(x_0)(\bar{x} - x_0)\| \\
&\le \frac{\mu L}{2}\|x_0 - \bar{x}\|^2 < c\|x_0 - \bar{x}\|^2(1 - \mu L\delta).
\end{aligned}$$

We apply the contraction mapping Theorem 5.4 with the following specifications:

$$\Phi = \Phi_0, \quad a = c\|x_0 - \bar{x}\|^2, \quad \gamma = \mu L\delta.$$

The preceding estimation gives us condition (a) in that theorem. To avoid using a twice in this theorem, let $r_0 = c\|x_0 - \bar{x}\|^2$. Now, to check condition (b), choose any $x, x' \in I\!B_{r_0}(\bar{x})$. Using (4) through (7) and the Lipschitz continuity of Df, we obtain

$$\begin{aligned}
&e(\Phi_0(x) \cap I\!B_{r_0}(\bar{x}), \Phi_0(x')) \le \\
&\quad e(G^{-1}(f(\bar{x}) + Df(\bar{x})(x - \bar{x}) - f(x_0) - Df(x_0)(x - x_0)) \cap I\!B_{r_0}(\bar{x}), \\
&\quad\quad G^{-1}(f(\bar{x}) + Df(\bar{x})(x' - \bar{x}) - f(x_0) - Df(x_0)(x' - x_0))) \\
&\quad \le \mu\|(Df(\bar{x}) - Df(x_0))(x - x')\| \le \mu L\delta\|x - x'\|.
\end{aligned}$$

Thus, condition (b) in Theorem 5.4 holds as well, and hence the mapping Φ_0 has a fixed point $x_1 \in I\!B_{r_0}(\bar{x})$; that is,

$$x_1 \in \Phi_0(x_1) = G^{-1}(f(\bar{x}) + Df(\bar{x})(x_1 - \bar{x}) - f(x_0) - Df(x_0)(x_1 - x_0)).$$

This yields that x_1 is obtained by way of (3) from x_0; moreover

$$\|x_1 - \bar{x}\| \le c\|x_0 - \bar{x}\|^2$$

and hence

$$\|x_1 - \bar{x}\| \le c\delta^2 < \delta.$$

Proceeding by induction, assume that there are iterates $x_2, \dots, x_k \in I\!B_\delta(\bar{x})$ of (3) that satisfy

$$\|x_i - \bar{x}\| \le c\|x_{i-1} - \bar{x}\|^2 \quad \text{for } i = 2, 3, \dots, k. \tag{14}$$

If $x_k = \bar{x}$ there is nothing more to prove. Let $x_k \ne \bar{x}$. By using the argument for Φ_0 in the above lines but this time for the mapping

$$x \mapsto \Phi_k(x) = G^{-1}(f(\bar{x}) + Df(\bar{x})(x - \bar{x}) - f(x_k) - Df(x_k)(x - x_k)),$$

we apply in the same manner Theorem 5.4 with

$$\Phi = \Phi_k, \quad a = c\|x_k - \bar{x}\|^2, \quad \lambda = \mu L\delta,$$

obtaining that there exists $x_{k+1} \in I\!B_\delta(\bar{x})$ satisfying (9) and (13). The induction step is complete and so is the proof. □

Proof of Theorem 15.2. According to Theorem 12.1 combined with Corollary 12.3, strong subregularity of $f + F$ at $\bar{x}$ for 0 is equivalent to isolated calmness of the inverse G^{-1} at 0 for $\bar{x}$ of the mapping G defined in (9); that is, there exist constants a, b, and μ such that

$$\|x - \bar{x}\| \leq \mu\|y - \bar{y}\| \quad \text{when} \quad x \in G^{-1}(y) \cap I\!B_a(\bar{x}) \text{ and } y \in I\!B_b(0). \tag{15}$$

Make a smaller if necessary such that $I\!B_a(\bar{x}) \subset O$, and let L be a Lipschitz constant of Df over $I\!B_a(\bar{x})$. Then choose $\delta > 0$ such that

$$\delta \leq \min\left\{a, \sqrt{\frac{2b}{L}}\right\}. \tag{16}$$

Using (12) and (16), we obtain that for every $x \in I\!B_\delta(\bar{x})$ we have

$$\|f(x) - f(\bar{x}) - Df(\bar{x})(x - \bar{x})\| \leq \frac{L}{2}\|x - \bar{x}\|^2 \leq \frac{1}{2}L\delta^2 < b. \tag{17}$$

Let $\{x_k\}$ be a sequence of points $x_1, \dots, x_k$ in $I\!B_\delta(\bar{x})$ obtained by the iteration (3) starting from a point $x_0 \in I\!B_\delta(\bar{x})$. For any k we have

$$f(\bar{x}) - f(x_k) + Df(x_k)(x_k - \bar{x}) \in f(\bar{x}) + Df(x_k)(x_{k+1} - \bar{x}) + F(x_{x+1}).$$

Hence, using (15), (16), and (17) we obtain

$$\|x_{k+1} - \bar{x}\| \leq \mu\|f(\bar{x}) - f(x_k) + Df(x_k)(x_k - \bar{x})\| \leq \frac{1}{2}\mu L\|x_k - \bar{x}\|^2.$$

This gives us the desired convergence rate.

In the case when the mapping $f + F$ is strongly regular, it is enough to mimic the proof of Theorem 15.1 where, instead Theorem 5.4, one applies the standard contraction mapping principle for functions given in Lecture 5. □

Proof of Theorem 15.3. We will first show that there exist constants α, β, and ℓ such that for every $A \in \bar{\partial} f(\bar{x})$, the mapping $I\!B_\beta(\bar{y}) \ni y \mapsto G_A^{-1}(y) \cap I\!B_\alpha(\bar{x})$ is a Lipschitz continuous function with Lipschitz constant ℓ.

Choose any $A \in \bar{\partial} f(\bar{x})$ as well as positive constants κ and κ' such that $\kappa' < \kappa$ and the mapping G_A^{-1} has a Lipschitz continuous localization at $\bar{y}$ for $\bar{x}$ with a Lipschitz constant κ' and neighborhoods $I\!B_a(\bar{x})$ and $I\!B_b(\bar{y})$. Make $b > 0$ smaller if necessary so that

$$b < 2a/\kappa'.$$

Let $\varepsilon > 0$ satisfy

$$\varepsilon < 1/\kappa', \quad \kappa'/(1 - \kappa'\varepsilon) \leq \kappa, \quad \varepsilon a \leq b/2 \text{ and } \kappa' b/2 \leq a(1 - \kappa'\varepsilon).$$

Choose any $A' \in \bar{\partial} f(\bar{x})$ such that $\|A - A'\| < \varepsilon$ and consider the mapping $G_{A'}$ associated with A'. For any $y \in I\!B_{b/2}(\bar{y})$ and any $x \in I\!B_a(\bar{x})$ we have

$$\|y + (A - A')(x - \bar{x}) - \bar{y}\| \le b/2 + \varepsilon a \le b.$$

Clearly,

$$x \in G_{A'}^{-1}(y) \cap I\!B_a(\bar{x}) \iff x \in \xi_y(x) := G_A^{-1}(y + (A - A')(x - \bar{x})) \cap I\!B_a(\bar{x}).$$

Let $y \in I\!B_{b/2}(\bar{y})$. We have

$$\|\xi_y(\bar{x}) - \bar{x}\| = \|G_A^{-1}(y) \cap I\!B_a(\bar{x}) - G_A^{-1}(\bar{y}) \cap I\!B_a(\bar{x})\| \le \kappa' b/2 \le a(1 - \kappa'\varepsilon).$$

Further, for any $x, x' \in I\!B_a(\bar{x})$ we obtain

$$\|\xi_y(x) - \xi_y(x')\| \le \kappa'\|(A - A')(x - x')\| \le \kappa'\varepsilon\|x - x'\|.$$

Thus, from the standard contraction mapping principle the function ξ_y has a unique fixed point $x(y)$ in $I\!B_a(\bar{x})$. Since $x(y) = G_{A'}^{-1}(y) \cap I\!B_a(\bar{x})$ for every $y \in I\!B_{b/2}(\bar{y})$, we conclude that the mapping $I\!B_{b/2}(\bar{y}) \ni y \mapsto x(y) = G_{A'}^{-1}(y) \cap I\!B_a(\bar{x})$ is single-valued. Furthermore, for any $y, y' \in I\!B_{b/2}(\bar{y})$ we have

$$\begin{aligned}\|x(y) - x(y')\| &= \|\xi_y(x(y)) - \xi_{y'}(x(y'))\| \\ &\le \kappa'\|y - y'\| + \kappa'\|(A - A')(x(y) - x(y'))\| \\ &\le \kappa'\|y - y'\| + \kappa'\varepsilon\|x(y) - x(y')\|,\end{aligned}$$

hence $x(\cdot)$ is Lipschitz continuous on $I\!B_{b/2}(\bar{y})$ with Lipschitz constant $\kappa'/(1-\kappa'\varepsilon) \le \kappa$. Thus, for a, b, ε, κ, and κ' as above, and for any $A' \in \bar{\partial} f(\bar{x})$ with $A' \in \operatorname{int} I\!B_\varepsilon(A)$, the mapping $G_{A'}^{-1}$ has a Lipschitz continuous single-valued localization at $\bar{y}$ for $\bar{x}$ with neighborhoods $I\!B_a(\bar{x})$ and $I\!B_{b/2}(\bar{y})$ and with Lipschitz constant κ. In other words, for any matrix A' which is sufficiently close to A the sizes of the neighborhoods and the Lipschitz constant associated with the localization $G_{A'}$ remain the same.

Pick any $A \in \bar{\partial} f(\bar{x})$ and corresponding a, b, and κ', and then $\varepsilon^A > 0$ and κ as required to obtain that for every $A' \in \operatorname{int} I\!B_{\varepsilon^A}(A)$ the mapping $I\!B_{b/2}(\bar{y}) \ni y \mapsto G_{A'}^{-1}(y) \cap I\!B_a(\bar{x})$ is a Lipschitz continuous function with Lipschitz constant κ. Since $\bar{\partial} f(\bar{x})$ is compact, from the open covering $\cup_{A \in \bar{\partial} f(\bar{x})} \operatorname{int} I\!B_{\varepsilon^A}(A)$ of $\bar{\partial} f(\bar{x})$ we can choose a finite subcovering with open balls $\operatorname{int} I\!B_{\varepsilon^{A_i}}(A_i)$; let a_i, b_i, and κ_i be the constants associated with the Lipschitz localizations for $G_{A_i}^{-1}$. Taking $\alpha = \min_i a_i$, $\beta = \min_i b_i/2$, and $\ell = \max_i \kappa_i$, we complete the proof of the statement given in the beginning of the proof.

Pick positive ε and a such that

$$\varepsilon < \frac{1}{2}\min\{1, 1/\ell\} \quad \text{and} \quad a \le \min\{\alpha, \beta\}. \tag{18}$$

Make a smaller if necessary so that, from the outer semicontinuity of the generalized Jacobian $\bar{\partial} f$, for any $x \in I\!B_a(\bar{x})$ and any $A \in \bar{\partial} f(x)$ there exists $\bar{A} \in \bar{\partial} f(\bar{x})$ such that

$$\|A - \bar{A}\| \le \varepsilon, \tag{19}$$

and also, from the semismoothness of f, for any $x \in I\!B_a(\bar{x})$ and any $A \in \bar{\partial} f(x)$,

$$\|f(x) - f(\bar{x}) - A(x - \bar{x})\| \leq \varepsilon \|x - \bar{x}\|. \tag{20}$$

Let $x_0 \in I\!B_a(\bar{x})$ and let $A_0 \in \bar{\partial} f(x_0)$. Choose $\bar{A}_0 \in \bar{\partial} f(\bar{x})$ such that, according to (19), $\|A_0 - \bar{A}_0\| \leq \varepsilon$. Then, for any $x \in I\!B_a(\bar{x})$, from (18), (19), and (20) we have

$$\|f(\bar{x}) - f(x_0) + A_0(x_0 - \bar{x}) + (\bar{A}_0 - A_0)(x - \bar{x})\| \leq \varepsilon a + \varepsilon a < a \leq \beta.$$

Consider the function

$$I\!B_a(\bar{x}) \ni x \mapsto \xi(x) := G_{\bar{A}_0}^{-1}(f(\bar{x}) - f(x_0) + A_0(x_0 - \bar{x}) + (\bar{A}_0 - A_0)(x - \bar{x})) \cap I\!B_a(\bar{x}).$$

Since

$$\xi(\bar{x}) = G_{\bar{A}_0}^{-1}(f(\bar{x}) - f(x_0) + A_0(x_0 - \bar{x})) \cap I\!B_a(\bar{x}) \quad \text{and} \quad \bar{x} = G_{\bar{A}_0}^{-1}(0) \cap I\!B_a(\bar{x}),$$

taking into account (18) and (20) we obtain that

$$\|\xi(\bar{x}) - \bar{x}\| \leq \ell \|f(\bar{x}) - f(x_0) + A_0(x_0 - \bar{x})\| \leq \ell \varepsilon a < (1 - \ell\varepsilon)a.$$

For any $x, x' \in I\!B_a(\bar{x})$ we have

$$\|\xi(x) - \xi(x')\| \leq \ell \|(\bar{A}_0 - A_0)(x - x')\| \leq \ell\varepsilon \|x - x'\|.$$

Hence, by the standard contraction mapping theorem for functions, there exists a unique $x_1 \in I\!B_a(\bar{x})$ such that $x_1 = \xi(x_1)$. This translates to the existence of a unique $x_1 \in I\!B_a(\bar{x})$ which satisfies the iteration (8) for $k = 0$. Furthermore, using (12), we obtain

$$\begin{aligned} \|x_1 - \bar{x}\| &= \|\xi(x_1) - \bar{x}\| \\ &= \|G_{\bar{A}_0}^{-1}(f(\bar{x}) - f(x_0) + A_0(x_0 - \bar{x}) \\ &\qquad +(\bar{A}_0 - A_0)(x_1 - \bar{x})) \cap I\!B_a(\bar{x}) - G_{\bar{A}_0}^{-1}(0) \cap I\!B_a(\bar{x})\| \\ &\leq \ell \|f(\bar{x}) - f(x_0) + A_0(x_0 - \bar{x})\| + \ell\varepsilon \|x_1 - \bar{x}\|, \end{aligned}$$

thus

$$\|x_1 - \bar{x}\| \leq \frac{\ell\varepsilon}{1 - \ell\varepsilon} \|x_0 - \bar{x}\|.$$

The induction step is completely analogous. Given $x_k \in I\!B_a(\bar{x})$ and $A_k \in \bar{\partial} f(x_k)$ we choose $\bar{A}_k \in \bar{\partial} f(\bar{x})$ such that $\|A_k - \bar{A}_k\| \leq \varepsilon$. Then, for the function

$$I\!B_a(\bar{x}) \ni x \mapsto \xi(x) := G_{\bar{A}_k}^{-1}(f(\bar{x}) - f(x_k) + A_k(x_k - \bar{x}) + (\bar{A}_k - A_k)(x - \bar{x})) \cap I\!B_a(\bar{x})$$

we show the existence of a unique fixed point x_{k+1} in $I\!B_a(\bar{x})$ satisfying

$$\|x_{k+1} - \bar{x}\| \leq \frac{\ell\varepsilon}{1 - \ell\varepsilon} \|x_k - \bar{x}\|.$$

Since the sequence $\{x_k\}$ is independent of the choice of ε and ε can be arbitrarily small, this establishes superlinear convergence of x_k to $\bar{x}$. Choosing $O = \mathbb{B}_a(\bar{x})$ we complete the proof. □

Lecture 16
The Constrained Linear-Quadratic Optimal Control Problem

In this and the following lectures, we switch gears to consider specific optimization problems in infinite-dimensional function spaces. In most of these lectures, we focus on a basic optimal control problem, which consists in minimizing a quadratic integral functional over a finite time interval, for a linear control system with control constrained to a closed and convex set. The problem is stated as follows:

$$\text{minimize} \quad \frac{1}{2}\int_0^1 x(t)^{\mathsf{T}} Q x(t) + u(t)^{\mathsf{T}} R u(t) dt \tag{1}$$

subject to

$$\dot{x}(t) = Ax(t) + Bu(t) \quad \text{for } t \in [0, 1], \qquad x(0) = x^0, \tag{2}$$

and

$$u(t) \in U \quad \text{for} \quad t \in [0, 1]. \tag{3}$$

Here the differential equation (2) is interpreted as a model of a *control system,* in which $x(t) \in I\!R^n$ is the so-called *state* of the system and $u(t) \in I\!R^m$ is the *control.* Both the state and control are functions of the scalar t which is interpreted as time, and $\dot{x}$ denotes the (usual) derivative of x with respect to t. It is assumed for simplicity that the time interval where the system (2) evolves is $[0, 1]$, but everything in this lecture remains valid for any finite time interval $[t_0, t_1]$ with $t_0 < t_1 < +\infty$. The state at time $t = 0$ called initial state, is fixed at x^0. The matrices A, B, Q, and R have dimensions fitting the dimensions of x and u, with Q and R assumed symmetric, Q positive semidefinite and R positive definite. The set $U \subset I\!R^m$ from which, according to (3), the values of the control $u(t)$ are to be selected, is assumed nonempty, closed, and convex. These are our *standing assumptions* for the optimal control problem (1)–(3). Note that the assumed positive definiteness of the symmetric matrix R is equivalent to the condition that there exists a positive constant μ such that

$$u^{\mathsf{T}} R u \geq \mu \|u\|^2 \quad \text{for all} \quad u \in I\!R^m. \tag{4}$$

© The Author(s), under exclusive license to Springer Nature Switzerland AG 2021

A. L. Dontchev, *Lectures on Variational Analysis*, Applied Mathematical Sciences 205, https://doi.org/10.1007/978-3-030-79911-3_16

Problem (1)–(3) is an optimization problem where the variables, state x and control u, are functions of time, the properties of which are still to be defined. In order to define the controls, we use the Hilbert space $L^2(I\!R^n, [0, 1])$, which consists of functions $u : [0, 1] \to I\!R^n$ that are Lebesgue integrable, equipped with the inner product

$$\langle u, v\rangle = \int_0^1 u(t)^\top v(t)dt,$$

and the corresponding norm

$$\|u\|_{L^2} = \sqrt{\langle u, u\rangle} = \left(\int_0^1 \|u(t)\|^2 dt\right)^{1/2}.$$

In the sequel we also use the short notation L^2 for $L^2(I\!R^l, [t_0, t_1])$, where the values of l, t_0, and t_1 depend on the context. In that case, the integral in (1) defining the objective function is in the sense of Lebesgue.

The function $t \mapsto x(t)$ the values of which represent the states is called *state trajectories*. The state trajectories are obtained as solutions of the differential equation (2) for controls $u \in L^2$. Since the control system is linear and the feasible controls are in L^2, the derivatives $\dot{x}$ are also elements of L^2. We denote the space of functions acting from $[0, 1]$ to $I\!R^n$ with derivatives in L^2 by $W^{1,2}(I\!R^n, [0, 1])$, or $W^{1,2}$ for short. When equipped with the inner product

$$\langle x, y\rangle_{W^{1,2}} = \langle x, y\rangle_{L^2} + \langle \dot{x}, \dot{y}\rangle_{L^2}$$

and the corresponding norm $\|x\|_{W^{1,2}} = \sqrt{\langle x, x\rangle_{W^{1,2}}}$, this is a Hilbert space.

Problem (1)–(3) is a control-constrained optimal control problem with a set of *feasible controls* defined as

$$\mathcal{U} = \{u \in L^2(I\!R^m, [0, 1]) \mid u(t) \in U \text{ for a.e. } t \in [0, 1]\}. \tag{5}$$

Note that the set $\mathcal{U}$ is a closed and convex set in L^2. Also note that any feasible control u, being an element of L^2, is actually an equivalence class of functions differing from each other on sets of measure zero in $[0, 1]$. That is, any feasible control u should have values $u(t)$ on any subset of the time interval $[0, 1]$ of full Lebesgue measure. In other words, we may change the values of a feasible control on any set of measure zero, e.g. on a set of finite, or even countable, number of points t without changing the feasibility and, more importantly, without changing the value of the objective function in (1) which measures optimality. Also, the constraints (2) and (3) are supposed to hold *almost everywhere* in $[0, 1]$, a.e. for short, which means that they are satisfied on a set of full measure 1.

In this and the following lectures, we will consider solutions of the initial value problem for a linear differential equation of the form

$$\dot{x}(t) = Ax(t) + w(t) \text{ for a.e. } t \in [t_0, t_1], \quad x(t_0) = x^0,$$

where $t_0 < t_1 < +\infty$, $x(t) \in I\!R^n$, A is an $n \times n$ matrix, and $w \in L^2(I\!R^n, [t_0, t_1])$. This equation has a unique solution which is in $W^{1,2}(I\!R^n, [t_1, t_2])$; moreover, this solution is given by the *Cauchy formula*:

$$x(t) = e^{A(t-t_0)}x^0 + \int_{t_0}^{t} e^{A(t-s)}w(s)ds \quad \text{for all } t \in [t_0, t_1],$$

where the exponent of a matrix is formally defined as the power series

$$e^A = \sum_{k=0}^{\infty} \frac{1}{k!}A^k.$$

There are (at least) two ways of looking at problem (1)–(3). We can think of it in terms of minimizing (1) over pairs (x, u) constrained by both (2) and (3), or we can regard x as a "dependent variable" produced from u through (2), and then reduce the problem to minimization with respect to u. Let us take the second approach. For any feasible control u, there is a unique state trajectory x specified by (2) and given by the Cauchy formula

$$x(t) = e^{At}x^0 + \int_0^t e^{A(t-\tau)}Bu(\tau)d\tau \quad \text{for all } t \in [0, 1]. \tag{6}$$

Define a mapping $T : L^2(I\!R^n, [0, 1]) \to L^2(I\!R^n, [0, 1])$ as follows:

$$(Tv)(t) = \int_0^t e^{A(t-\tau)}v(\tau)d\tau \quad \text{for a.e. } t \in [0, 1]. \tag{7}$$

This mapping is linear and bounded. Also, with a slight abuse of notation, denote by B the mapping from $L^2(I\!R^m, [0, 1])$ to $L^2(I\!R^n, [0, 1])$ associated with the matrix B and defined as $(Bu)(t) = Bu(t)$; later we will do the same for the matrices Q and R. Then the formula for x in (6) comes out as

$$x(t) = e^{At}x^0 + (TB)(u)(t). \tag{8}$$

According to the definition of T, we are treating the state trajectory x as a function of time belonging to $L^2(I\!R^n, [0, 1])$. In fact, the state trajectory is an element of the smaller space $W^{1,2}(I\!R^n, [0, 1])$. Also, it will be convenient to use for the state an intermediate space, the Banach space $C(I\!R^n; [0, 1])$ of continuous functions over the interval $[0, 1]$ with values in $I\!R^n$ equipped with the norm

$$\|x\|_C = \max_{t \in [0,1]} \|x(t)\|.$$

By changing the order of integration, we obtain

$$\begin{aligned}\langle x, Tv\rangle &= \int_0^1 x(t)^\mathsf{T} \int_0^t e^{A(t-\tau)}v(\tau)d\tau dt \\ &= \int_0^1 \int_\tau^1 x(t)^\mathsf{T} e^{A(t-\tau)}dt\, v(\tau)d\tau = \langle T^* x, v\rangle.\end{aligned}$$

Hence, the adjoint map T^* to T is given by

$$(T^*x)(t) = \int_t^1 e^{A^\top(\tau-t)}x(\tau)d\tau \quad \text{for a.e. } t \in [0, 1].$$

Furthermore, $(TB)^* = B^*T^*$, where B^* is the mapping acting from $L^2(I\!R^n, [0, 1])$ to $L^2(I\!R^m, [0, 1])$ and associated with the transposed matrix $B^\top$. Specifically,

$$((TB)^*x)(t) = \int_t^1 B^\top e^{A^\top(\tau-t)}x(\tau)d\tau \quad \text{for a.e. } t \in [0, 1].$$

As we did for the mapping B, we consider Q and R as linear bounded mappings acting between L^2 spaces, that is, $(Qx)(t) = Qx(t)$ and $(Ru)(t) = Ru(t)$. Let $v(t) = e^{At}x^0$. With this notation, problem (1)–(3) can be written in the form

$$\text{minimize } \frac{1}{2}(\langle x, Qx\rangle + \langle u, Ru\rangle) \quad \text{subject to} \quad x = v + TBu, \quad u \in \mathcal{U}. \tag{9}$$

By substituting $x = v + TBu$ in the cost function, we convert (9) to a problem of minimizing a modified objective function which depends only on the control u. Defining the bounded linear mapping

$$\mathcal{A} = B^*T^*QTB + R,$$

which acts from L^2 to itself, and $w = B^*T^*Qv$, the substitution reduces problem (9) to

$$\text{minimize } \frac{1}{2}\langle u, \mathcal{A}u\rangle + \langle w, u\rangle \quad \text{subject to } u \in \mathcal{U}, \tag{10}$$

where we drop the constant term $\langle v, Qv\rangle/2$ which does not affect optimality. Since Q and R are symmetric, the mapping $\mathcal{A}$ is self-adjoint, $\mathcal{A}^* = \mathcal{A}$. The set $\mathcal{U}$ in (5) is a closed and convex subset of L^2. Because of the bound (4), the mapping $\mathcal{A}$ satisfies the condition

$$\langle u, \mathcal{A}u\rangle \geq \mu\|u\|_{L^2}^2 \quad \text{for all } u \in \mathcal{U} - \mathcal{U}, \tag{11}$$

where the constant μ is the same as the one in (4). Note that (11) is the coercivity condition introduced in Lecture 9. This condition implies that the quadratic form in (10) is strongly convex relative to $\mathcal{U}$. Thus, problem (10) is an optimization problem in the space L^2 with a strongly convex objective function minimized over a closed and convex set.

In Lecture 11, we considered a quadratic optimization problem in a general Hilbert space. Specifically, from Theorem 11.1 specialized to the Hilbert space L^2, we obtain that under condition (11), problem (10) has a solution which is unique; moreover, a necessary and sufficient optimality condition for problem (10) is represented by the variational inequality:

$$w + \mathcal{A}u + N_{\mathcal{U}}(u) \ni 0. \tag{12}$$

Let us reiterate that the optimal control, as a function of time, is an element of an equivalence class of functions differing from each other only on sets of measure zero in $[0, 1]$. Later in this lecture, we will go on to show that one can pick a particular function from the equivalence class which has better continuity properties with respect to time.

We will now derive a more specific form of the optimality system (12). Let

$$\psi = T^* Qx,$$

where x solves (6) for a given control u. Then, through the Cauchy formula, ψ is given by

$$\psi(t) = \int_t^1 e^{A^\mathsf{T}(\tau - t)}(Qx)(\tau)d\tau.$$

Hence, ψ is a continuous function which is differentiable almost everywhere in $[0, 1]$ and its derivative $\dot{\psi}$ is in L^2. Furthermore, ψ satisfies

$$\dot{\psi}(t) = -A^\mathsf{T}\psi(t) - Qx(t) \quad \text{for a.e. } t \in [0, 1], \quad \psi(1) = 0. \tag{13}$$

The vector $\psi(t)$ is called the *adjoint* or *co-state* variable and (13) is called the *adjoint equation*. Note that, since x is a continuous function of t whose derivative is in L^2, the adjoint trajectory $t \mapsto \psi(t)$ is twice differentiable almost everywhere in $[0, 1]$ and has its second time derivative in L^2.

Bearing in mind that

$$\psi = T^* Qx = T^*(QTBu + v) \text{ and } w = B^* T^* Qv,$$

we can re-express the variational inequality (12) in terms of ψ as

$$\langle Ru + B^*\psi, y - u\rangle \geq 0 \qquad \text{for all} \quad y \in \mathcal{U}. \tag{14}$$

Recall that B^* stands for the linear mapping associated with the transpose B^T of the matrix B.

We need next a standard fact from Lebesgue integration. For a function φ on $[0, 1]$, a point $\hat{t} \in (0, 1)$ is said to be a *Lebesgue point* of φ when

$$\lim_{\varepsilon \to 0} \frac{1}{2\varepsilon} \int_{\hat{t}-\varepsilon}^{\hat{t}+\varepsilon} \varphi(\tau)d\tau = \varphi(\hat{t}).$$

It is known that when φ is integrable on $[0, 1]$, its set of Lebesgue points is of full measure, in this case measure one.

Now, let u be the optimal control and let x and ψ be the associated optimal trajectory and the adjoint trajectory, respectively. Let $\hat{t}$ be a Lebesgue point of u (remember that the set of such $\hat{t}$ is of full measure). Pick any $w \in U$, and for $0 < \varepsilon < \min\{\hat{t}, 1 - \hat{t}\}$ consider the function

$$\hat{u}_\varepsilon(t) = \begin{cases} w & \text{for } t \in (\hat{t} - \varepsilon, \hat{t} + \varepsilon), \\ u(t) & \text{otherwise.} \end{cases}$$

Then for every sufficiently small ε the function $\hat{u}_\varepsilon$ is a feasible control, i.e., belongs to the set $\mathcal{U}$ in (5), and from (14) we obtain

$$\int_{\hat{t}-\varepsilon}^{\hat{t}+\varepsilon} (Ru(\tau) + B^\mathsf{T}\psi(\tau))^\mathsf{T}(w - u(\tau))d\tau \geq 0.$$

Since $\hat{t}$ is a Lebesgue point of the function under the integral (we know that ψ is continuous and hence its set of Lebesgue points is the entire interval $[0, 1]$), we can pass to zero with ε and by taking into account that $\hat{t}$ is an arbitrary point from a set of full measure in $[0, 1]$ and that w can be any element of U, come to the following *pointwise* variational inequality which holds for a.e. $t \in [0, 1]$:

$$(Ru(t) + B^\mathsf{T}\psi(t))^\mathsf{T}(w - u(t)) \geq 0 \quad \text{for all } w \in U. \tag{15}$$

As is easily seen, (15) implies (14) as well, and hence these two variational inequalities defined in different spaces are equivalent.

Summarizing, we can now say that the state-control pair $(\bar{x}, \bar{u})$ is the unique solution of problem (1)–(3) with corresponding adjoint trajectory $\bar{\psi}$ if and only if the triple $(\bar{x}, \bar{u}, \bar{\psi})$ solves the following boundary value problem coupled with a pointwise variational inequality:

$$\begin{cases} \dot{x}(t) = Ax(t) + Bu(t), & x(0) = x^0, \\ \dot{\psi}(t) = -A^\mathsf{T}\psi(t) - Qx(t), & \psi(1) = 0, \end{cases} \tag{16a}$$

$$0 \in Ru(t) + B^\mathsf{T}\psi(t) + N_U(u(t)) \quad \text{for a.e. } t \in [0, 1]. \tag{16b}$$

Recall that the set $\mathcal{U}$ of feasible controls defined in (5) is a subset of the space L^2 of square integrable functions. Each function of that space is actually an equivalence class of functions that differ from each other sets in $[0, 1]$ of measure zero. That is, for any representative of the optimal control $\bar{u}$ there exists a set of full measure in $[0, 1]$ such that (16a,b) holds for every t in this set. We will now show that in the equivalence class of optimal control functions $\bar{u}$, there exists an optimal control which is continuously differentiable with respect to t on $[0, 1]$ and which satisfies (16b) *for all* $t \in [0, 1]$.

As a solution of a linear differential equation the right side of which is a continuous function, the adjoint trajectory $\bar{\psi}$ is continuously differentiable, hence Lipschitz continuous, in $t \in [0, 1]$. Let u be any representative function of $\bar{u}$ such that (16b) holds for all $t \in \sigma$ where σ is a set of full measure in $[0, 1]$. For a fixed $t \notin \sigma$ we define $u(t)$ to be the unique solution of the following strongly monotone variational inequality in $\mathbb{R}^m$:

$$q(t) \in Ru + N_U(u), \quad \text{where } q(t) = -B^\mathsf{T}\bar{\psi}(t). \tag{17}$$

Then the resulting control u is within the equivalence class $\bar{u}$ of optimal controls, and, moreover, the vector $u(t)$ satisfies (16b) for all $t \in [0, 1]$ with matrix R being positive definite. Noting that q is a Lipschitz continuous function in t on $[0, 1]$, we get from Theorem 10.1 that for each fixed $t \in [0, 1]$ the solution mapping of (17) is Lipschitz continuous with respect to $q(t)$. Since the composition of Lipschitz continuous functions is Lipschitz continuous, the particular optimal control function u which satisfies (16a,b) for all $t \in [0, 1]$ is Lipschitz continuous in t on $[0, 1]$. Observe that changing the values of the control on a set of measure zero does not affect the optimal trajectory $\bar{x}$ nor the adjoint trajectory $\bar{\psi}$, and that (16a) can be assumed to hold for all $t \in [0, 1]$. The following theorem summarizes what we obtained so far for the optimal control problem (1)–(3):

Theorem 16.1. *There exists a unique optimal solution $(\bar{x}, \bar{u})$ of problem (1)–(3) together with a corresponding unique adjoint trajectory $\bar{\psi}$ such that the triple $(\bar{x}, \bar{u}, \bar{\psi})$ is a solution of the optimality system (16a,b). Moreover, from the equivalence class of optimal control functions, there exists an optimal control for which (16b) holds for all $t \in [0, 1]$ and which is Lipschitz continuous with respect to t in $[0, 1]$.*

The function

$$H(x, \psi, u) = \frac{1}{2}(x^{\mathsf{T}} Qx + u^{\mathsf{T}} Ru) + \psi^{\mathsf{T}}(Ax + Bu)$$

is called the *Hamiltonian* associated with the problem (1)–(3). Noting that (16b) is a necessary and sufficient optimality condition for minimizing $H(x, \psi, u)$ with respect to u over the set U, the optimality system (16a,b) can be written as

$$\begin{cases} \dot{x}(t) = \nabla_{\psi} H(x(t), \psi(t), u(t)), & x(0) = a, \\ \dot{\psi}(t) = -\nabla_x H(x(t), \psi(t), u(t)), & \psi(1) = 0, \\ H(x(t), \psi(t), u(t)) = \min_{v \in U} H(x(t), \psi(t), v) \end{cases} \tag{18}$$

for a.e. $t \in [0, 1]$. First-order optimality conditions of this kind can be obtained for much more general optimal control problems than the one considered here; these are commonly known as the *Pontryagin principle*. In the following lecture, we will state the Pontryagin principle for a nonlinear optimal control problem. We just proved that principle for the linear-quadratic problem (1)–(3).

We consider next a modification of problem (1)–(3) in which the objective/cost function includes also a quadratic term of the final state $x(1)$:

$$\text{minimize } J(x, u) = \frac{1}{2}\left[x(1)^{\mathsf{T}} F x(1) + \int_0^1 \left(x(t)^{\mathsf{T}} Qx(t) + u(t)^{\mathsf{T}} Ru(t)\right) dt\right] \tag{19}$$

subject to (2) and (3), where, in addition to the standing assumptions, F is assumed to be a symmetric and positive semidefinite matrix. In this case, the Pontryagin principle needs the following modification: The final condition for the adjoint variable becomes

$$\psi(1) = -Fx(1). \tag{20}$$

We will now prove that this is indeed the case.

In the beginning of this lecture, we mentioned that we can think of problem (1)–(3) in terms of minimizing (1) over pairs (x, u) constrained by both (2) and (3), or we can regard x as a "dependent variable" obtained from u through (2). We then used the second approach to deal with the problem. We will now handle the modified problem to minimize (19) subject to (2),(3) by using the first approach. In this case we treat the differential equation (2) as an equality constraint, to which apply the Lagrange principle discussed in Lecture 1 (around formulas (9) and (10)) for a finite-dimensional problem, but an extension of the general idea to infinite dimensions is straightforward. We give next a bit less rigorous derivation of the optimality conditions by using "variations" instead of derivatives, which is actually closer to the original idea to characterize optimality in calculus of variations.

Introduce the Lagrange function

$$L(x, u, \psi) = J(x, u) + \int_0^1 \psi^{\mathsf{T}}(t)(-\dot{x}(t) + Ax(t) + Bu(t))dt,$$

where $J(x, u)$ is as in (19). According to the Lagrange principle, we minimize the Lagrange function with respect to (x, u) with x in L^2 and u in the set $\mathcal{U}$ given in (5). Let $(\bar{x}, \bar{u})$ be a solution of the problem and consider a variation $(\bar{x} + \delta x, \bar{u} + \delta u) \in L^2$, where $\bar{u} + \delta u \in \mathcal{U}$ and $\delta x(0) = 0$ because $x(0)$ is fixed. The deviation of the Lagrange function from its minimal value is

$$\begin{aligned} 0 &\le L(\bar{x} + \delta x, \bar{u} + \delta u, \psi) - L(\bar{x}, \bar{u}, \psi) \\ &= \delta x(1)^{\mathsf{T}} F x(1) + \int_0^1 \left(\delta x(t)^{\mathsf{T}} Q\bar{x}(t) + \delta u(t)^{\mathsf{T}} R\bar{u}(t)\right) dt \\ &\quad + \int_0^1 \psi^{\mathsf{T}}(t)(-\delta\dot{x}(t) + A\delta x(t) + B\delta u(t))dt. \end{aligned}$$

Using integration by part for the term involving $\delta\dot{x}$ and rearranging the other terms gives us

$$\begin{aligned} 0 &\le L(\bar{x} + \delta x, \bar{u} + \delta u, \psi) - L(\bar{x}, \bar{u}, \psi) \\ &= \delta x(1)^{\mathsf{T}}(\psi(1) + F\bar{x}(1)) + \int_0^1 \delta x(t)^{\mathsf{T}}(\dot{\psi}(t) + A^{\mathsf{T}}\psi(t) + Q\bar{x}(t))dt \\ &\quad + \int_0^1 \delta u(t)^{\mathsf{T}}(B^{\mathsf{T}}\psi(t) + R\bar{u}(t))dt, \end{aligned}$$

where we take into account that $\delta x(0) = 0$. Hence, the variation of the Lagrange function with respect to x will be zero when ψ solves the adjoint equation (13) with the final condition (20). We see that the adjoint variable defined as the solution of (13),(20) is the Lagrange multiplier associated with the equality constraints representing the state equation. The variation of the Lagrange function with respect to u satisfies

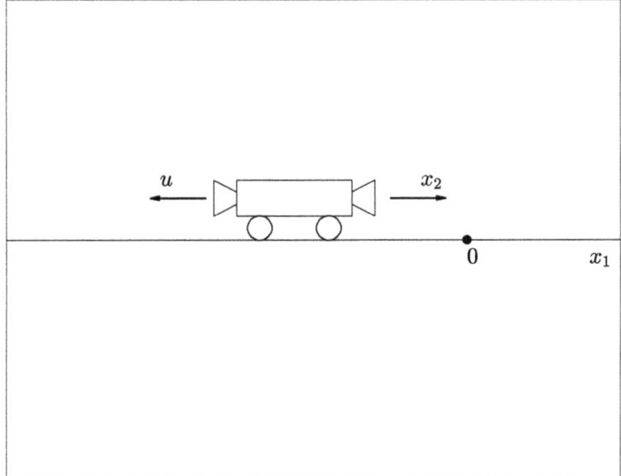

Fig. 16.1: The rocket car

$$0 \leq \int_0^1 \delta u(t)^\top (B^\top \psi(t) + R\bar{u}(t))dt.$$

This last condition is of course the same as (14), hence we end up with the variational inequality (16b). The above derivation can be also used to show that when the initial state $x(0)$ is not fixed but "free," then the initial value $\psi(0)$ of the adjoint variable must be taken zero. Symmetrically, if the final state $x(1)$ is fixed, then $\psi(1)$ must be free.

The Rocket Car/Double Integrator Example. We will illustrate the above analysis with a classical example in control. Consider the motion of a car equipped with two rocket engines, one in front and one on the back (Fig. 16.1). Using the real line as a coordinate system, let us denote the position of the car at time t by $x_1(t)$ and its velocity by $x_2(t)$. The motion of the car is described by the second law of Newton

$$ma(t) = F(t),$$

where m is the mass of the car, $a(t)$ is the acceleration at time t, and $F(t)$ is the rocket thrust at time t which is used as control. For simplicity, assume that the mass of the car $m = 1$. Then, inasmuch as the time derivative $\dot{x}_1(t)$ is the velocity $x_2(t)$, and the time derivative $\dot{x}_2(t)$ of the acceleration is the thrust $u(t)$, we obtain the following model of the car:

$$\begin{aligned} \dot{x}_1(t) &= x_2(t), \ x_1(0) = x_1^0, \\ \dot{x}_2(t) &= \ u(t), \ \ x_2(0) = x_2^0. \end{aligned} \tag{21}$$

This model is also called the *double integrator* in the literature. It has a two-dimensional state $x = (x_1, x_2)$ and one-dimensional control u. The corresponding Cauchy formula has the form

$$x(t) = e^{At}x^0 + \int_0^t e^{A(t-s)}bu(s)ds,$$

where

$$A = \begin{pmatrix} 0 & 1 \\ 0 & 0 \end{pmatrix}, \quad b = \begin{pmatrix} 0 \\ 1 \end{pmatrix}.$$

Integrating directly (21), we obtain

$$e^{At} = \begin{pmatrix} 1 & t \\ 0 & 1 \end{pmatrix}.$$

The state evolution in time is then described by

$$x_1(t) = x_1^0 + x_2^0 t + \int_0^t (t-s)u(s)ds,$$
$$x_2(t) = x_2^0 + \int_0^t u(s)ds.$$

Consider now the problem to optimally control the car over the time interval [0, 1] so that the objective function

$$0.5\left(x_1^2(1) + x_2^2(1) + \int_0^1 u^2(t)dt\right)$$

is minimized. The Hamiltonian has the form

$$H(x, \psi, u) = 0.5u^2 + \psi_1 x_2 + \psi_2 u$$

and the adjoint equation becomes

$$\begin{aligned} \dot{\psi}_1(t) &= \quad 0, \\ \dot{\psi}_2(t) &= -\psi_1(t) \end{aligned} \tag{22}$$

with the end-time condition, according to (20),

$$\begin{aligned} \psi_1(1) &= -x_1(1) \\ \psi_2(1) &= -x_2(1). \end{aligned}$$

Thus, from (22) the adjoint variable ψ_1 is constant in time while ψ_2 is linear in time. There are no constraints for the control, hence the derivative of H with respect to u must be zero, which gives

$$u(t) = -\psi_2(t).$$

Hence, the optimal control is a linear function in time, while the optimal velocity x_2 and position x_1 are, respectively, a quadratic polynomial and a cubic polynomial.

Lecture 17
Regularity in Nonlinear Control

In this lecture we consider a control system described by a nonlinear ordinary differential equation of the form

$$\dot{x}(t) = g(x(t), u(t)), \quad x(0) = x^0, \tag{1}$$

over the interval $[0, 1]$. Here, as for the linear-quadratic problem in the preceding lecture, $x(t) \in I\!R^n$ is the state of the system, while $u(t) \in I\!R^m$ is the control, both at time t. We assume that the function $g : I\!R^n \times I\!R^m \to I\!R^n$ is continuously differentiable everywhere.

Both state and control are functions of time t over the interval $[0, 1]$. In order to complete the description of the control system (1), we have to state what kind of functions of time they are to be; that is, to define the set of feasible controls and state trajectories. For the linear-quadratic problem in the preceding lecture, the spaces of feasible state and control functions are chosen to be the Hilbert spaces $W^{1,2}$ and L^2, respectively. The space L^2 works well for linear systems but is not suitable for nonlinear ones, mainly because differentiating the function g in (1) considered as acting from $L^2 \times L^2$ to L^2 would need additional assumptions such as for example global Lipschitz continuity; such an assumption would exclude polynomials of higher degree. Therefore we consider a smaller set of feasible controls consisting of functions $u : [0, 1] \to I\!R^m$ that are Lebesgue measurable and essentially bounded over $[0, 1]$. When equipped with the norm

$$||u||_{L^\infty} = \text{ess sup}_{t \in [0,1]} ||u(t)||,$$

the set of such functions is a Banach space denoted by $L^\infty(I\!R^m, [0, 1])$. The essential supremum "ess sup" here is defined as

$$\text{ess sup}_{t \in [0,1]} \mu(t) = \inf\{ \sup_{t \in [0,1]} \nu(t) \mid \mu(t) = \nu(t) \text{ for a.e. } t \in [0, 1]\}.$$

© The Author(s), under exclusive license to Springer Nature Switzerland AG 2021
A. L. Dontchev, *Lectures on Variational Analysis*, Applied Mathematical Sciences 205, https://doi.org/10.1007/978-3-030-79911-3_17

When the space of values and the t-interval are clear from the context, we use the short notation L^∞. The state trajectory x, which is a solution of the initial value problem (1) for a given control function u, is assumed to belong to the space of continuous functions x with values in $I\!R^n$ that are differentiable almost everywhere and whose derivatives are in L^∞. When equipped with the norm

$$\|x\|_{W^{1,\infty}} = \|x\|_{L^\infty} + \|\dot{x}\|_{L^\infty},$$

this space is a Banach space; we denote it by $W^{1,\infty}(I\!R^n, [0, 1])$, or $W^{1,\infty}$ for short. The differential equation (1) is considered to hold for almost every $t \in [0, 1]$, where, as before, *almost every* (a.e.) means that this relation holds for t in a subset of $[0, 1]$ with full Lebesgue measure. In these spaces differentiating the function g is not a problem; to obtain the state trajectory control derivatives of g, respectively, from $W^{1,\infty}$ and L^∞ to L^∞, it is sufficient to take the derivatives of g with respect to x and u in $I\!R^n$ and $I\!R^m$ and then define the resulting expressions in the respective function spaces. Later in the book we also employ for state trajectories the Banach space $C(I\!R^n; [0, 1])$ of continuous functions equipped with the maximum norm

$$\|x\|_C = \max_{t\in[0,1]} \|x(t)\|.$$

The choice of spaces for controls and states is chosen not only for mathematical convenience, which is a factor, but also because of practical considerations. Another function space we will use for the controls is the space of Lebesgue integrable functions on the interval $[0, 1]$ with values in $I\!R^m$ equipped with the norm

$$\|u\|_{L^1} = \int_0^1 \|u(t)\|dt.$$

This Banach space is denoted by $L^1(I\!R^m, [0, 1])$, or simply by L^1. The corresponding space of state trajectories is the Banach space $W^{1,1}$ of continuous functions with derivatives in $L^1(I\!R^n, [0, 1])$. Of course, all these spaces can be defined on an interval $[a, b]$, finite or infinite, in the real line.

A simple change of variables reduces system (1) to one with initial state zero, $x(0) = 0$; therefore, without loss of generality, we assume that $x(0) = x^0 = 0$. It is convenient for our further analysis to "implant" the zero initial condition into the space of state trajectories, by assuming that x belongs to

$$W_0^{1,\infty}(I\!R^n, [0, 1]) = \{x \in W^{1,\infty}(I\!R^n, [0, 1]),\ x(0) = 0\},$$

which is a closed linear subspace of the space $W^{1,\infty}(I\!R^n, [0, 1])$.

In the first part of this lecture we consider system (1) subject to pointwise in time inequality constraints on the state and control. Namely, given a function $G : I\!R^{n+m} \longrightarrow I\!R^l$ that we assume continuously differentiable everywhere, we define the set of feasible states and controls as the set of pairs (x, u) that belong to respective spaces of functions and satisfy system (1); that is, x is a solution of (1) for u, and (x, u) satisfies the state-control inequality constraint

$$G(x(t), u(t)) \le 0 \quad \text{for a.e. } t \in [0, 1]. \tag{2}$$

System (1) under the constraint (2) can be identified with a generalized equation for the variable $(x, u) \in W_0^{1,\infty} \times L^\infty$, of the form

$$0 \in f(x, u) + F(x, u),$$

where

$$f(x, u) = \begin{pmatrix} \dot{x} - g(x, u) \\ G(x, u) \end{pmatrix} \quad \text{and} \quad F(x, u) = \begin{pmatrix} 0 \\ L_+^\infty \end{pmatrix}.$$

Here $f : W_0^{1,\infty}(I\!R^n, [0, 1]) \times L^\infty(I\!R^n, [0, 1]) \to L^\infty(I\!R^l, [0, 1])$ and F acts from $W_0^{1,\infty}(I\!R^n, [0, 1]) \times L^\infty(I\!R^m, [0, 1])$ to the product of the set consisting of just the zero function in $L^\infty(I\!R^n, [0, 1])$ and the set L_+^∞ defined as

$$L_+^\infty = \{w \in L^\infty(I\!R^l, [0, 1]) \mid w(t) \ge 0 \ \text{for a.e. } t \in [0, 1]\}.$$

Then $(f + F)^{-1}(0)$ is the set of solutions of (1)–(2) or the set of *feasible* state-control pairs (x, u).

We will now focus on metric regularity of the mapping $f + F$. The proof of the following result utilizes Corollary 5.3 of the (extended Lyusternik–Graves) Theorem 5.2 as well as (Robinson–Ursescu) Theorem 6.5, providing a characterization of metric regularity of the feasible set mapping in terms of the linearization of (1)–(2), involving the matrices

$$A = \nabla_x g(\bar{x}, \bar{u}), \ B = \nabla_u g(\bar{x}, \bar{u}), \ S = \nabla_x G(\bar{x}, \bar{u}), \ K = \nabla_u G(\bar{x}, \bar{u}).$$

Theorem 17.1. *The mapping $f + F$ described above is metrically regular at $(\bar{x}, \bar{u})$ for 0 if and only if there exist a constant $\alpha > 0$, and functions $w \in W_0^{1,\infty}$ and $v \in L^\infty$ such that, for a.e. $t \in [0, 1]$,*

$$\begin{cases} \dot{w}(t) = A(t)w(t) + B(t)v(t), \\ [G(\bar{x}(t), \bar{u}(t)) + S(t)w(t) + K(t)v(t)]_i \le -\alpha \ \text{for } i = 1, 2, \ldots, l. \end{cases}$$

Proof. According to Corollary 5.3, metric regularity of the mapping $f + F$ at the reference point $(\bar{x}, \bar{u})$ for 0 is equivalent to metric regularity at $(\bar{x}, \bar{u})$ for 0 of the linearized mapping

$$(x, u) \mapsto \begin{pmatrix} \dot{x} - \dot{\bar{x}} - A(x - \bar{x}) - B(u - \bar{u}) \\ G(\bar{x}, \bar{u}) + S(x - \bar{x}) + K(u - \bar{u}) \end{pmatrix} + \begin{pmatrix} 0 \\ L_+^\infty \end{pmatrix}, \tag{3}$$

where we take into account that $\dot{\bar{x}} = g(\bar{x}, \bar{u})$ and the differentiability of the function f with derivative represented in an obvious way by the mapping composed of the matrices A, B, S, and K defined before the theorem. This mapping has closed and convex graph; hence, we can apply (Robinson–Ursescu) Theorem 6.5 according to which metric regularity of the mapping (3) is equivalent to the existence of $\delta > 0$ such that for any $y = (p, q)$ with $\|y\|_{L^\infty} \le \delta$, the following system has a solution $(x, u) \in W_0^{1,\infty} \times L^\infty$ for a.e. $t \in [0, 1]$:

$$\begin{aligned}&\dot{x}(t) - \dot{\bar{x}}(t) = A(t)(x(t) - \bar{x}(t)) + B(t)(u(t) - \bar{u}(t)) + p(t),\\&G(\bar{x}(t), \bar{u}(t)) + S(t)(x(t) - \bar{x}(t)) + K(t)(u(t) - \bar{u}(t)) + q(t) \le 0.\end{aligned} \tag{4}$$

Taking $p = 0$, $q = (\alpha, \dots, \alpha)$, and then $w = x - \bar{x}$, $v = u - \bar{u}$, this last condition (4) implies the condition in the statement of the theorem. Conversely, let (w, v) satisfy the condition in the statement for some $\alpha > 0$, let x be the solution of the differential equation in (4) corresponding to the control $u = v + \bar{u}$ and $x(0) = 0$, and let $y = (p, q)$ be a function in L^∞. Note that $x = w + \bar{x} + \Lambda p$, where Λ is a bounded linear mapping from L^∞ to $W_0^{1,\infty}$; specifically, $\tilde{x} = \Lambda p$ is the solution of the linear equation $\dot{\tilde{x}} = A\tilde{x} + p$, $\tilde{x}(0) = 0$. Hence,

$$\begin{aligned}&G(\bar{x}, \bar{u}) + S(x - \bar{x}) + K(u - \bar{u}) + q\\&\qquad = G(\bar{x}, \bar{u}) + Kv + S(w + \Lambda p) + q \le -\bar{\alpha} + S\Lambda p + q,\end{aligned}$$

where $\bar{\alpha} = (\alpha, \dots, \alpha) \in I\!R^l_\infty$ with $\alpha > 0$. Clearly, for any $y = (p, q) \in L^\infty$ with a sufficiently small norm, the right side of the last inequality will be nonpositive. This completes the proof. □

We consider next the following nonlinear optimal control problem:

$$\text{minimize} \quad \int_0^1 \varphi(x(t), u(t))dt \tag{5}$$

subject to

$$\dot{x}(t) = g(x(t), u(t)), \quad u(t) \in U \;\text{ for a.e. } t \in [0, 1],$$
$$x \in W_0^{1,\infty}(I\!R^n), \quad u \in L^\infty(I\!R^m),$$

where the functions $\varphi : I\!R^n \times I\!R^m \to I\!R$ and $g : I\!R^n \times I\!R^m \to I\!R^n$ are twice continuously differentiable everywhere and U is a convex and closed set in $I\!R^m$. Dealing with constraints for the state x would considerably complicate the analysis and will not be considered here.

Without going into much detail, we present next a necessary optimality condition for problem (5) known as the Pontryagin principle; we derived a specific form of it for the linear-quadratic problem considered in the preceding lecture. Let $(\bar{x}, \bar{u})$ be a solution of (5). Define the Hamiltonian

$$H(x, u, \psi) = \varphi(x, u) + \psi^{\mathsf{T}} g(x, u).$$

The Pontryagin principle says that there exists a function $\bar{\psi} \in W^{1,\infty}(I\!R^n)$ called the *adjoint* or *co-state* variable, such that the triple $\bar{\xi} := (\bar{x}, \bar{\psi}, \bar{u})$ is a solution of the following two-point boundary value problem coupled with a pointwise variational inequality:

$$\begin{cases}\dot{x}(t) = g(x(t), u(t)), \quad x(0) = 0,\\ \dot{\psi}(t) = -\nabla_x H(x(t), u(t), \psi(t)), \quad \psi(1) = 0,\\ 0 \in \nabla_u H(x(t), u(t), \psi(t)) + N_U(u(t)) \quad \text{for a.e. } t \in [0, 1],\end{cases} \tag{6}$$

where $N_U(u)$ is the normal cone to the set U at the point u.

We will now rewrite problem (6) as a generalized equation. First, in analogy with the space $W_0^{1,\infty}$ of the state trajectories, let us consider the adjoint variables in the subspace

$$W_1^{1,\infty}(I\!R^n,[0,1]) = \{\psi \in W^{1,\infty}(I\!R^n,[0,1]),\ \psi(1)=0\}.$$

Introduce the spaces

$$\begin{aligned} X &= W_0^{1,\infty}(I\!R^n,[0,1]) \times W_1^{1,\infty}(I\!R^n,[0,1]) \times L^\infty(I\!R^m,[0,1]), \\ Y &= L^\infty(I\!R^n,[0,1]) \times L^\infty(I\!R^n,[0,1]) \times L^\infty(I\!R^m,[0,1]). \end{aligned} \tag{7}$$

Further, define the mappings

$$(x,\psi,u) \mapsto h(x,\psi,u) = \begin{pmatrix} \dot{x} - g(x,u) \\ \dot{\psi} + \nabla_x H(x,u,\psi) \\ \nabla_u H(x,u,\psi) \end{pmatrix}$$

and

$$(x,\psi,u) \mapsto \mathcal{H}(x,\psi,u) = \begin{pmatrix} 0 \\ 0 \\ \mathcal{V}(u) \end{pmatrix}$$

both acting from X to Y, the first being a function and the second a set-valued mapping, where

$$\mathcal{V}(u) = \{v \in L^\infty(I\!R^m,[0,1]) \mid 0 \in v(t) + N_U(u(t)) \text{ for a.e. } t \in [0,1]\}.$$

Then, in terms of $\xi = (x,\psi,u)$, the necessary optimality system (6) can be written as the generalized equation

$$\text{find } \xi \in X \text{ such that } h(\xi) + \mathcal{H}(\xi) \ni 0.$$

It is important to note that the latter relation is *not* a variational inequality since Y is not the space dual to X, at least.

We will now focus on strong regularity of the mapping $h + \mathcal{H}$ by applying Corollary 8.7 of the (extended Robinson) Theorem 8.5. First, observe that the assumed twice differentiability of φ and g implies that the function h is Fréchet differentiable in the spaces X and Y defined in (7). Then, according to Corollary 8.7, strong regularity of the mapping $h + \mathcal{H}$ at $\bar{\xi}$ for 0 is equivalent to strong regularity of its linearization

$$\xi \mapsto h(\bar{\xi}) + Dh(\bar{\xi})(\xi - \bar{\xi}) + \mathcal{H}(\xi)$$

at $\bar{\xi}$ for 0, where $Dh(\bar{\xi})$ is the derivative of h at $\bar{\xi} = (\bar{x},\bar{\psi},\bar{u})$. By changing the variables from ξ to $\xi - \bar{\xi}$ and going back to the original notation, the mapping representing the linearization of $h + \mathcal{H}$ becomes

$$(x, \psi, u) \mapsto \begin{pmatrix} \dot{x} - Ax - Bu \\ \dot{\psi} + A^{\mathsf{T}}\psi + Qx + Du \\ \nabla_x H(\bar{x}, \bar{\psi}, \bar{u}) + Ru + D^{\mathsf{T}}x + B^{\mathsf{T}}\psi \end{pmatrix} + \begin{pmatrix} 0 \\ 0 \\ \mathcal{V}(u + \bar{u}) \end{pmatrix}. \tag{8}$$

Here the matrices A, B, Q, D, and R are the second derivatives of the Hamiltonian H with respect to ψ, x, and u at $(\bar{x}, \bar{\psi}, \bar{u})$, respectively; that is,

$$A = \nabla_x \bar{g} = \nabla_{\psi x}\bar{H},\ B = \nabla_u \bar{g} = \nabla_{\psi u}\bar{H},\ Q = \nabla_{xx}\bar{H},\ D = \nabla_{xu}\bar{H},\ R = \nabla_{uu}\bar{H},$$

where the bar indicates that the derivatives are evaluated at the optimal $(\bar{x}, \bar{\psi}, \bar{u})$. Note that all these matrices depend on time t. We now focus on strong regularity of the mapping (8) at 0 for 0, by employing the following coercivity-type assumption:

(A1) There exists a positive constant μ such that

$$\begin{pmatrix} x^{\mathsf{T}} & u^{\mathsf{T}} \end{pmatrix} \begin{pmatrix} Q(t) & D(t) \\ D(t)^{\mathsf{T}} & R(t) \end{pmatrix} \begin{pmatrix} x \\ u \end{pmatrix} \geq \mu \|u\|^2 \text{ for all } (x, u) \in \mathbb{R}^n \times (U - U) \text{ and a.e. } t \in [0, 1].$$

According to the definition, strong regularity of the mapping (8) at 0 for 0 means the following: the solution mapping of the perturbed system

$$\begin{cases} \dot{x} - Ax - Bu = p, \quad x(0) = 0, \\ \dot{\psi} + A^{\mathsf{T}}\psi + Qx + Du = s, \quad \psi(1) = 0, \\ \nabla_x H(\bar{x}, \bar{\psi}, \bar{u}) + Ru + D^{\mathsf{T}}x + B^{\mathsf{T}}\psi + \mathcal{V}(u + \bar{u}) \ni r \end{cases} \tag{9}$$

has a single-valued localization around 0 for 0 relative to the spaces X and Y in (7), which is Lipschitz continuous with respect to the parameter (p, s, r) around zero.

Along with system (9) consider the linear-quadratic problem

$$\text{minimize} \quad \int_0^1 \frac{1}{2}\left(x^{\mathsf{T}}Qx + 2x^{\mathsf{T}}Du + u^{\mathsf{T}}Ru\right) - x^{\mathsf{T}}s - u^{\mathsf{T}}r\, dt \tag{10}$$

subject to

$$\dot{x} = Ax + Bu + p, \quad x(0) = 0, \quad u \in \mathcal{U}, \tag{11}$$

where the set of feasible control functions now is

$$\mathcal{U} = \{u \in L^2(\mathbb{R}^m, [0, 1]) \mid u(t) \in U + \bar{u}(t) \text{ for a.e. } t \in [0, 1]\}.$$

Changing at this point the space of controls from L^∞ to L^2 is for convenience only, without restricting the generality of the analysis, which allows us to apply the argument used in the preceding Lecture 16. Following that argument, the next step is to eliminate the state variable from problem (10)–(11). To do that, we have to take into account that the matrices A and B now depend on time t, in which case in the Cauchy formula and in the definition of the mapping T in Lecture 16 the matrix exponent should be replaced by the so-called fundamental matrix solution of the linear differential equation in (11). This is a function Φ acting from $\mathbb{R}^2$ to the set of $n \times n$ matrices and defined as the solution of the differential equation

$$\frac{d}{dt}\Phi(t,\tau) = A(t)\Phi(t,\tau), \quad \Phi(\tau,\tau) = I,$$

where I is the $n \times n$ identity matrix. When the matrix A is independent of t, then $\Phi(t,\tau) = e^{A(t-\tau)}$ and we come to the Cauchy formula in Lecture 16.

Now, define the mapping $T : L^2(I\!R^n, [0,1]) \to L^2(I\!R^n, [0,1])$ as

$$(Tw)(t) = \int_0^t \Phi(t,\tau)w(\tau)d\tau \quad \text{for a.e. } t \in [0,1].$$

The adjoint T^* to T can be found in a way completely analogous to finding the adjoint to T in the preceding Lecture 16. Denote by B the mapping acting from L^2 to L^2 and associated with the time-dependent matrix $B(t)$, and the same for the matrices Q, D, and R. Utilizing the adjoint T^*, define the self-adjoint bounded linear mapping $\mathcal{A} : L^2 \to L^2$ by

$$\mathcal{A} = B^*T^*QTB + 2D^*TB + R$$

and the perturbation function

$$w(z) = -r - B^*T^*s + (2D^* + B^*T^*Q)Tp \quad \text{for } z = (p,s,r).$$

Then, problem (10)–(11) is written as

$$\text{minimize } \langle w(z), u\rangle + \frac{1}{2}\langle u, \mathcal{A}u\rangle \text{ subject to } u \in \mathcal{U}. \tag{12}$$

Observe that the assumption (A1) implies the coercivity condition

$$\langle u, \mathcal{A}u\rangle \geq \mu\|u\|_2^2 \quad \text{for all } u \in \mathcal{U} - \mathcal{U}. \tag{13}$$

Furthermore, under condition (A1), the objective function in (10) is convex over $\mathcal{U}$; hence problem (10)–(11) is equivalent to the variational inequality

$$w(z) + \mathcal{A}u + N_{\mathcal{U}}(u) \ni 0. \tag{14}$$

We come to the realm of Lecture 16, where we considered a perturbed quadratic program and the associated variational inequality representing the associated optimality system. Taking into account that $w(z)$ is an affine function of $z = (p,s,r)$ and using the argument in the proof of Theorem 16.1 with some minor adjustments, we obtain that for each z problem (12), or equivalently (14), has a unique solution $u[z]$ and, moreover, the function $z \mapsto u[z]$ is globally Lipschitz continuous from $L^2(I\!R^n \times I\!R^n \times I\!R^m, [0,1])$ to $L^2(I\!R^m, [0,1])$. Here we use square brackets to indicate that u is a function of the function z. For any $z = (p,s,r)$ in L^2, let $x[u,z]$ be the corresponding solution of the state equation (11) for u and p. For any $z, z' \in L^2$, taking norms in the difference of the corresponding solutions of (11) expressed by the Cauchy formula, we obtain that, for some constant $c > 0$,

$$\|x[u[z],z](t) - x[u[z'],z'](t)\| \leq c(\|u[z] - u[z']\|_2 + \|p - p'\|_2).$$

Taking the supremum over $[0, 1]$ on the left and using the Lipschitz continuity of the function $z \mapsto u[z]$ on the right, we conclude that the optimal trajectory mapping $z \mapsto x[z]$ is Lipschitz continuous from the L^2 space for z to the space $C(I\!R^n, [0, 1])$ for x. Using the same argument for the adjoint trajectory ψ, we get that $z \mapsto \psi[z]$ is Lipschitz continuous from the L^2 to $C(I\!R^n, [0, 1])$ as well. Hence, the mapping $z = (p, s, r) \mapsto (x[z], \psi[z], u[z])$ describing the solution of system (9) is single-valued and Lipschitz continuous from the space $L^2(I\!R^n \times I\!R^n \times I\!R^m, [0, 1])$ to the space $C(I\!R^n \times I\!R^n, [0, 1]) \times L^2(I\!R^m, [0, 1])$.

There is more to be said about the Lipschitz continuity of the solution of the linearized optimality system (9) if we assume that

(A2) the matrices B, D, and R are Lipschitz continuous in $[0, 1]$.

Under (A2) and the additional assumption that the perturbation function r is Lipschitz continuous on $[0, 1]$, and using the fact that for a fixed perturbation z the perturbed state $x[z]$ and co-state $\psi[z]$ are Lipschitz continuous in $[0, 1]$ as well, we repeat the argument around formula (17) in Lecture 16 with $q(t) = r(t) - B^{\mathsf{T}}(t)\psi[z](t) - D^{\mathsf{T}}(t)x[z](t)$ to obtain that from the equivalence class of optimal controls for problem (10), and there exists an optimal control being a function $t \mapsto u(z)(t)$ of time $t \in [0, 1]$ for which the pointwise in time variational inequality in (9) holds for all $t \in [0, 1]$ and which is Lipschitz continuous with respect to t on $[0, 1]$. Moreover, the argument used in the proof of Theorem 16.1 applied to the quadratic problem equivalent to that variational inequality gives us that its solution mapping is Lipschitz continuous with respect to z, in this case from the space $L^2(I\!R^n \times I\!R^n, [0, 1]) \times C(I\!R^m, [0, 1])$ to the space $C(I\!R^m, [0, 1])$. Summarizing, we have:

Theorem 17.2. *Suppose that the condition (A1) holds. Then the solution mapping of system* (9) *is single-valued and Lipschitz continuous from the space* $L^2(I\!R^n \times I\!R^n \times I\!R^m, [0, 1])$ *to the space* $C(I\!R^n \times I\!R^n, [0, 1]) \times L^2(I\!R^m, [0, 1])$. *In addition, if condition (A2) is satisfied and each perturbation* $z = (p, s, r)$ *is such that* r *is a Lipschitz continuous function in* $[0, 1]$, *then the corresponding optimal control* $u[z]$ *is Lipschitz continuous in time* $t \in [0, 1]$, *while the solution mapping* $z \mapsto (x[z], \psi[z], u[z])$ *is Lipschitz continuous from the space* $L^2(I\!R^n \times I\!R^n, [0, 1]) \times C(I\!R^m, [0, 1])$ *to* $C(I\!R^n \times I\!R^n \times I\!R^m, [0, 1])$. *Hence, under these conditions, the mapping representing the Pontryagin principle* (6) *for the nonlinear problem* (5) *is strongly regular at the respective function spaces for the state, co-state, and control.*

According to (extended Robinson) Theorem 8.5, after corresponding adjustments, these conditions are also sufficient for Lipschitz stability of the following perturbed version of problem (5):

$$\text{minimize} \int_0^1 \varphi(x(t), u(t), \sigma(t))dt \tag{15}$$

subject to

$$\dot{x}(t) = g(x(t), u(t), \sigma(t)), \quad u(t) \in U \text{ for a.e. } t \in [0, 1],$$

where σ is a time-dependent parameter. We will not go into this further; relevant references are included in the bibliographical remarks at the end of the book.

The following example illustrates the case when the coercivity condition is violated.

Example. Consider the problem

$$\min \int_0^2 -x(t) + u(t)dt$$

subject to

$$\dot{x}(t) = u(t),\ x(0) = 0, \quad u(t) \in [0, 1],\ t \in [0, 2].$$

The feasible control functions are assumed to belong to the space L^1, while the feasible state trajectories are in $W^{1,1}$. The optimal control $\bar{u}$ is unique and given by

$$\bar{u}(t) = \begin{cases} 1 & \text{for } t \in [0, 1], \\ 0 & \text{for } t \in [1, 2]. \end{cases}$$

Consider the perturbed problem

$$\min \int_0^2 (-x(t) + (1 - r(t))u(t))dt$$

subject to the same constraints, where the perturbation parameter $r \in L^\infty$. The optimal control of the perturbed problem has the form

$$u[r](t) = \begin{cases} 1 & \text{for } t \in [0, 1 + r(t)], \\ 0 & \text{for } t \in (1 + r(t), 2]. \end{cases}$$

First, observe that the optimal control $u[r]$ is Lipschitz continuous as a mapping from L^∞ to L^1 but not Lipschitz continuous as from L^∞ to L^2. The optimality system of the perturbed problem has the form

$$\begin{cases} \dot{x}(t) - u(t) = 0, \quad x(0) = 0, \\ \dot{\psi}(t) = 1, \quad \psi(2) = 0, \\ \psi(t) + 1 - r(t) + N_{[0,1]}(u(t)) \ni 0. \end{cases} \tag{16}$$

Consider the perturbation

$$r(t) = \begin{cases} 1 - t & \text{for } t \in [1, 1 + \varepsilon], \\ 0 & \text{otherwise}, \end{cases},$$

where $\varepsilon \in (0, 1)$. Noting that $\psi(t) = -2 + t$, the variational inequality in (16) becomes

$$N_{[0,1]}(u(t)) \ni 0 \text{ for } t \in [1, 1 + \varepsilon],$$

which has infinitely many solutions. Noting that the perturbation r can have arbitrarily small L^∞ norm when ε is chosen sufficiently small, we obtain that the mapping describing the optimality system of the original problem (for $r = 0$) cannot be strongly regular as a mapping acting from L^∞. Of course this does not contradict the Lipschitz continuity of the optimal control $u[r]$ since not every solution of the optimality system is optimal. But there is more in this example. Observe that the optimality system has a unique solution for $r = 0$ and the L^1 norm of the difference between the solutions for any r and $r = 0$ is proportional to the L^∞ norm of r. Taking into account that perturbing the other relations in (16) does not matter; this leads to the conclusion that the mapping describing the optimality system of the original problem (for $r = 0$) is strongly subregular. It turns out that this situation is actually typical for problems linear in control, see the bibliographical remarks to this lecture.

Lecture 18
Discrete Approximations

In this lecture we focus on the issue of solving optimal control problems numerically. This is a problem of recovering a function of time, in this case an optimal control, which is determined implicitly as a solution of the problem. Aside from very special cases where a solution can be derived analytically, finding an optimal control means producing numerically an approximation of it. To obtain such an approximation, the usual way is to approximate the problem by another problem that we can solve numerically. Such an approximation would be justified if the solution of the approximate problem is an acceptable approximation of the solution of the original problem. This is a part of the broader issue to determine when an approximation of a problem gives an approximation of a solution to that problem. We start with the linear-quadratic optimal control problem considered in Lecture 16. Then we deal with the nonlinear optimal control problem studied in the preceding lecture.

Consider the linear-quadratic optimal control problem

$$\text{minimize} \quad \frac{1}{2}\int_0^1 x(t)^\mathsf{T} Q x(t) + u(t)^\mathsf{T} R u(t) dt \tag{1}$$

subject to

$$\begin{aligned} &\dot{x}(t) = Ax(t) + Bu(t) \quad \text{for a.e. } t \in [0,1], \qquad x(0) = x^0, \\ &u(t) \in U \quad \text{for a.e. } t \in [0,1], \quad x \in W^{1,2}(I\!R^n, [0,1]),\ u \in L^2(I\!R^m, [0,1], \end{aligned} \tag{2}$$

under the standing assumptions stated in Lecture 16; namely, the matrices Q and R are symmetric, Q is positive semidefinite, R is positive definite, and U is a nonempty, closed, and convex set in $I\!R^m$. The control functions belong to the space L^2 of square integrable functions over $[0, 1]$, in our case with values in the set U. As solutions of the linear ordinary differential equation in (2) with constant coefficients and control functions in the right side, any state trajectory x is an element of $W^{1,2}$, that is, a continuous function of $t \in [0, 1]$ with values in $I\!R^n$ whose derivative $\dot{x}$ is in L^2.

© The Author(s), under exclusive license to Springer Nature Switzerland AG 2021

A. L. Dontchev, *Lectures on Variational Analysis*, Applied Mathematical Sciences 205, https://doi.org/10.1007/978-3-030-79911-3_18

We know from Theorem 16.1 that the optimal control can be regarded as a Lipschitz continuous function in time $t \in [0, 1]$ and the optimal state, control, and the adjoint variable satisfy the following optimality system for all $t \in [0, 1]$:

$$\begin{cases} \dot{x}(t) = Ax(t) + Bu(t), \quad x(0) = x^0, \\ \dot{\psi}(t) = -A^{\mathsf{T}}\psi(t) - Qx(t), \quad \psi(1) = 0, \\ 0 \in Ru(t) + B^{\mathsf{T}}\psi(t) + N_U(u(t)). \end{cases} \tag{3}$$

Also recall that under the standing assumptions, system (3) is equivalent to problem (1)–(2) in the sense that the unique solution $(\bar{x}, \bar{u})$ of (1)–(2) coupled with the solution $\bar{\psi}$ of the adjoint equation for $\bar{x}$ solves the system (3), and, conversely, if $(\bar{x}, \bar{u}, \bar{\psi})$ solves (3), then $(\bar{x}, \bar{u})$ is the unique solution of (1)–(2).

At the current stage of matters it is safe to assume that we can solve numerically finite-dimensional optimization problems in a satisfactory way; therefore, we may choose to compute an approximate solution of problem (1)–(2) by converting it into a finite-dimensional problem. To do that, we need to choose first a finite-dimensional representation of functions on $[0, 1]$. Suppose that the interval $[0, 1]$ is divided into N pieces $[t_i, t_{i+1}]$, $i = 0, 1, \ldots, N-1$, by, for simplicity, equally spaced nodes t_i, $i = 0, 1, \ldots, N$, with $t_0 = 0$ and $t_N = 1$, with mesh size $h = t_i - t_{i-1} = 1/N$. To approximate the control we employ piecewise constant functions across the grid $\{t_i\}$ that are continuous from the right at each t_i, $i = 0, 1, \ldots, N-1$ and from the left at $t_N = 1$. That is, for a given N, we consider a subset of L^2 consisting of piecewise constant functions, given by

$$\Big\{ u \,\Big|\, u(t) = u(t_i) \text{ for } t \in [t_i, t_{i+1}),\ i = 0, 1, \ldots, N-2, \\ u(t) = u(t_{N-1}) \text{ for } t \in [t_{N-1}, t_N] \Big\}.$$

We also need to employ finite-dimensional approximations of the operators of integration and differentiation involved in problem (1)–(2). For the integral in the cost function in (1) we use the simplest rectangular formula of the integral and for the state equation in (2) the corresponding Euler scheme over the mesh $\{t_i\}$. In this way we obtain the following discrete-time problem:

$$\text{minimize} \quad J^N(x, u) = \frac{1}{2} \sum_{i=0}^{N-1} h \left(x_i^{\mathsf{T}} Q x_i + u_i^{\mathsf{T}} R u_i \right) \tag{4}$$

subject to

$$\begin{aligned} x_{i+1} &= (I + hA)x_i + hBu_i, \quad x_0 = x^0, \\ u_i &\in U \quad \text{for } i = 0, 1, \ldots, N-1, \end{aligned} \tag{5}$$

where I is the $n{\times}n$ identity matrix. This is a finite-dimensional optimization problem, the solution of which depends on the mesh size h. On the standing assumptions, problem (4)–(5) is a convex optimization problem.

Another way to approach the approximation of problem (1)–(2) is to invoke a discretization of its optimality system (3). The Euler scheme applied to (3), forward for the state equation and backward for the adjoint equation, combined with restricting the variational inequality to the nodes t_i, results in a discrete-time boundary value problem coupled with a finite-dimensional variational inequality, of the form:

$$\begin{cases} x_{i+1} = (I + hA)x_i + hBu_i, \quad x_0 = x^0, \\ \psi_i = (I + hA^{\mathsf{T}})\psi_{i+1} + hQx_{i+1}, \quad \psi_N = 0, \qquad \text{for } i = 0, 1, \ldots, N-1. \\ 0 \in Ru_i + B^{\mathsf{T}}\psi_i + N_U(u_i), \end{cases} \tag{6}$$

In this and the following lectures we use the following standard result:

Discrete Gronwall Inequality. *Consider nonnegative reals α, α_i, $i = 0, \ldots, N$, and β that satisfy*

$$\alpha_0 \le \alpha \;\; \textit{and} \;\; 0 \le \alpha_{i+1} \le \alpha + \beta \sum_{j=0}^{i} \alpha_j \;\; \textit{for } i = 0, \ldots, N.$$

Then

$$\alpha_i \le \alpha(1+\beta)^i \;\; \textit{for } i = 0, \ldots, N.$$

Similarly, if

$$\alpha_N \le \alpha \;\; \textit{and} \;\; 0 \le \alpha_{i+1} \le \alpha + \beta \sum_{j=i+1}^{N} \alpha_j \;\; \textit{for } i = 0, \ldots, N,$$

then

$$\alpha_i \le \alpha(1+\beta)^{N-i} \;\; \textit{for } i = 0, \ldots, N.$$

Theorem 18.1. *There exists a solution $(\bar{x}^N, \bar{u}^N)$ of problem (4)–(5) and this solution is unique; moreover, $(\bar{x}^N, \bar{u}^N)$ is the unique solution of (4)–(5) if and only if the solution $\bar{\psi}^N$ of the discrete-time adjoint equation for $\bar{x}^N$ is such that the triple $(\bar{x}^N, \bar{u}^N, \bar{\psi}^N)$ is a solution of system (6).*

Proof. We are in finite dimensions, which makes showing existence simple, by applying the standard argument used, e.g., to prove the Weierstrass theorem. For a fixed N consider a minimizing sequence (x^k, u^k) for (4)–(5). By using the coercivity of the objective function it is easy to show that this sequence is bounded. Indeed, utilizing the assumed positive definiteness of R and the positive semidefiniteness of Q, we have

$$\mu \sum_{i=0}^{N-1} h\|u_i^k\|^2 \le J^N(x^k, u^k),$$

where μ is the constant in the coercivity condition (4) in Lecture 16. Since the right side of this inequality converges to the optimal value that is finite, we obtain that the sequence u^k is bounded (remember that N is fixed). By applying the discrete Gronwall inequality we conclude that the sequence x^k is bounded as well. Hence, the minimizing sequence has a convergent subsequence, the limit of which will belong to the feasible set since this set is closed. The objective function is quadratic, hence continuous; thus, the limit will be a solution of (4)–(5). The strict convexity of the objective function makes the solution unique.

In order to derive the optimality condition, we apply the Lagrange multiplier rule by treating the state equation as linear equality constraints. Introduce the Lagrange function

$$L^N(x, u, \psi) = J^N(x, u) + \sum_{i=0}^{N-1} \psi_i^\top (x_{i+1} - (I + hA)x_i - hBu_i).$$

According to the optimality condition given in Lecture 1, formula (10), the first-order necessary optimality condition, which is also sufficient because of convexity, has the form

$$\nabla_x L^N(x, u, \psi) = 0$$
$$\nabla_u L^N(x, u, \psi) + N_U(u) \ni 0,$$

where $N_U(u)$ is the normal cone mapping to the set U at the point $u \in U$. Note that the constraint qualification is satisfied in this case since the equality constraint is represented by the linear state equation. Differentiating with respect to u is straightforward:

$$\nabla_{u_i} L^N(x, u, \psi) = hRu_i + hB^\top \psi_i.$$

The differentiation with respect to x is slightly more involved. First, note that x_0 is fixed and not involved in the minimization; hence, we do not differentiate with respect to it. Differentiating with respect to x_i, $i = 1, \ldots, N-1$, we obtain

$$\nabla_{x_i} L^N(x, u, \psi) = hQx_i + \psi_{i-1} - (I + hA)\psi_i,$$

while differentiating with respect to x_N gives us

$$\nabla_{x_N} L^N(x, u, \psi) = \psi_{N-1}.$$

By having the derivatives equal to zero and shifting notation from $(\psi_0, \ldots, \psi_{N-1})$ to $(\psi_1, \ldots, \psi_N)$ we obtain the adjoint equation in the form required. □

Suppose that for each given N we can solve problem (4)–(5), or equivalently (6), exactly, obtaining vectors $u_i^N \in U$, $i = 0, \ldots, N-1$ as well as $x_i^N \in I\!R^n$, and $\psi_i^N \in I\!R^n$, for $i = 0, \ldots, N$. Then a solution (x^N, u^N, ψ^N) is identified with a function on $[0, 1]$, where x^N and ψ^N are the piecewise linear and continuous interpolations across the grid $\{t_i\}$ over $[0, 1]$ of the points $(x^0, x_1^N, \ldots, x_N^N)$ and $(\psi_0^N, \psi_1^N, \ldots, \psi_{N-1}^N, 0)$, respectively, and u^N is the piecewise constant interpolation of $(u_0^N, u_1^N, \ldots, u_{N-1}^N)$ that is continuous from the right across the grid points $t_i = ih$, $i = 0, 1, \ldots, N-1$ and from the left at $t_N = 1$. The functions x^N and ψ^N are piecewise linear and continuous and their derivatives $\dot{x}^N$ and $\dot{\psi}^N$ are piecewise constant functions that have the same continuity properties in t as the control u^N.

Thus, (u^N, x^N, ψ^N) is a function defined in the whole interval $[0, 1]$. For the discrete-time controls u^N we employ the L^∞ norm we used in Lecture 17, and for the corresponding state and adjoint trajectories we use the uniform (maximum) norm $\|x\|_C$ of the space of continuous functions $C(I\!R^n; [0, 1])$. The following theorem gives an error estimate for the discrete approximation considered:

Theorem 18.2. *Consider problem* (1)–(2) *under the standing assumptions and let* $(\bar{u}, \bar{x}, \bar{\psi})$ *be the solution of the optimality system* (3) *for all* $t \in [0, 1]$ *with the optimal control* $\bar{u}$ *Lipschitz continuous in time* t *on* $[0, 1]$. *For* $N \in \{1, 2 \dots\}$, *consider also the discretization* (4)–(5) *with mesh size* $h = 1/N$, *and the associated optimality system* (6), *and denote by* $(\bar{u}^N, \bar{x}^N, \bar{\psi}^N)$ *its unique solution extended by interpolation to the interval* $[0, 1]$ *in the manner described above. Then there exists a constant* c *such that for all* $N = 1, 2, \dots$ *the following estimate holds:*

$$\|\bar{u}^N - \bar{u}\|_{L^\infty} + \|\bar{x}^N - \bar{x}\|_C + \|\bar{\psi}^N - \bar{\psi}\|_C \le ch.$$

This result implies that, by choosing a sufficiently fine discretization, we can approximate the solution of problem (1)–(2) with any required accuracy; furthermore, the error of the discretization is bounded by a constant times the step-size h, which, according to the terminology in numerical analysis, makes it a first-order approximation. This error bound is sharp for the Euler scheme and hence for the discrete approximation (4)–(5) of problem (1)–(2). Using higher-order schemes may improve the order of approximation, but this may require more regular behavior of the optimal control as a function of time; see the bibliographical remarks. A proof of Theorem 18.2 is presented at the end of the lecture.

In preparation for the developments in the next two lectures, we consider an embedding of the discretized problem (4)–(5) into a family of problems, where the initial time is t_k for some $k \in \{0, 1, \dots, N-1\}$ and the initial state x^0 is treated as a parameter $y \in I\!R^n$. That is, instead of (4)–(5), we deal with the problem

$$\text{minimize} \quad J^N(x, u) = \frac{1}{2} \sum_{i=k}^{N-1} h \left(x_i^{\mathsf{T}} Q x_i + u_i^{\mathsf{T}} R u_i \right) \tag{7}$$

subject to

$$\begin{aligned} &x_{i+1} = (I + hA)x_i + hBu_i, \quad x_k = y, \\ &u_i \in U \quad \text{for } i = k, k+1, \dots, N-1. \end{aligned} \tag{8}$$

The system representing the first-order necessary optimality condition has the form

$$\begin{cases} x_{i+1} = (I + hA)x_i + hBu_i, \quad x_k = y, \\ \psi_i = (I + hA^{\mathsf{T}})\psi_{i+1} + hQx_{i+1}, \quad \psi_N = 0, \\ 0 \in Ru_i + B^{\mathsf{T}}\psi_i + N_U(u_i) \end{cases} \tag{9}$$

for $i = k, k+1, \dots, N-1$.

The following theorem is an extension of Theorems 18.1 and 18.2 to the setting of problem (7)–(8). In its statement we call *data* of problem (1)–(2) the matrices

A, B, Q, and R, the set U, and the initial condition x^0. We also use the following notation: For $k \in \{0, 1, \ldots, N-1\}$ and $y \in I\!R^n$, $\bar{u}[t_k, y]$ is the optimal control for the continuous-time problem (1)–(2) on the interval $[t_k, 1]$ with the initial state condition $x(t_k) = y$, which is Lipschitz continuous in time t on $[t_k, 1]$, and $\bar{x}[t_k, y]$ and $\bar{\psi}[t_k, y]$ are the corresponding continuous-time optimal state and adjoint trajectories, so that $(\bar{u}[t_k, y], \bar{x}[t_k, y], \bar{\psi}[t_k, y])$ solves the optimality system (3) for all $t \in [t_k, 1]$.

Theorem 18.3. *For any natural number N, any $k \in \{0, 1, \ldots, N-1\}$, and any $y \in I\!R^n$, the discrete-time problem* (7)–(8) *has a solution that is unique; moreover, $\bar{u}^N[k, y]$ is the optimal control and $\bar{x}^N[k, y]$ is the optimal state trajectory for* (7)–(8) *if and only if there is a solution of the adjoint equation such that the triple $(\bar{x}^N[k, y], \bar{\psi}^N[k, y], \bar{u}^N[k, y])$ is the solution of the optimality system* (9). *Furthermore, for any $b > 0$ and any $y \in I\!B_b(x^0)$ there exists a positive constant c dependent only on the data of the problem* (1)–(2) *and b such that the following estimate holds:*

$$\max_{k \le i \le N-1} \Big((\|\bar{x}^N[k, y]_i - \bar{x}[k, y](t_i)\| + \|\bar{\psi}^N[k, y]_i - \bar{\psi}[k, y](t_i)\| + \|\bar{u}^N[k, y]_i - \bar{u}[k, y](t_i)\| \Big) \le ch.$$

The proof of this theorem basically repeats the proof of Theorem 18.2 with some modifications that allow to monitor the dependence of the constant c on k and b. It is left as an exercise supplied with a guide.

Exercise 18.4. *Prove Theorem 18.3.*

Guide. As in the proof of Theorem 18.2 given at the end of the lecture, first show boundedness of the optimal control $\bar{u}^N[k, y]$ in L^2 by a constant that is independent of N, k, and $y \in I\!B_b(x^0)$. Then the discrete Gronwall inequality gives uniform boundedness of $\bar{x}^N[k, y]$ and $\bar{\psi}^N[k, y]$, which in turn yields uniform boundedness of $\bar{u}^N[k, y]$. The next step is to derive an estimate analogous to (7) with the new initial time and state also showing that the corresponding Lipschitz constants depend on the data of the problem and b but not on the choice of N and k. This is straightforward by just checking the steps of the proof of Theorem 18.2. □

In the second part of this lecture we study a discrete approximation to the nonlinear optimal control problem from the preceding Lecture 17:

$$\text{minimize} \quad \int_0^1 \varphi(x(t), u(t))dt \tag{10}$$

subject to

$$\begin{aligned} &\dot{x}(t) = g(x(t), u(t)), \quad u(t) \in U \text{ for a.e. } t \in [0, 1], \\ &x \in W_0^{1,\infty}(I\!R^n), \quad u \in L^\infty(I\!R^m), \end{aligned} \tag{11}$$

where the functions $\varphi : \mathbb{R}^n \times \mathbb{R}^m \to \mathbb{R}$ and $g : \mathbb{R}^n \times \mathbb{R}^m \to \mathbb{R}^n$ are twice continuously differentiable everywhere and U is a convex and closed set in $\mathbb{R}^m$. As before, we denote $W_0^{1,\infty} = \{x \in W^{1,\infty} \mid x(0) = 0\}$. Recall that, in terms of the Hamiltonian

$$H(x, u, \psi) = \varphi(x, u) + \psi^{\mathsf{T}} g(x, u),$$

the optimality system of problem (10)–(11) has the form

$$\begin{cases} \dot{x}(t) &= g(x(t), u(t)), \quad x(0) = 0, \\ \dot{\psi}(t) &= -\nabla_x H(x(t), u(t), \psi(t)), \quad \psi(1) = 0, \\ 0 &\in \nabla_u H(x(t), u(t), \psi(t)) + N_U(u(t)). \end{cases} \tag{12}$$

As in Lecture 17, introduce the spaces $X = W_0^{1,\infty}(\mathbb{R}^n, [0,1]) \times W_1^{1,\infty}(\mathbb{R}^n, [0,1]) \times L^\infty(\mathbb{R}^m, [0,1])$ and $Y = L^\infty(\mathbb{R}^n], [0,1]) \times L^\infty(\mathbb{R}^n, [0,1]) \times L^\infty(\mathbb{R}^m, [0,1])$ that correspond to the domain and the range of the mapping appearing in (12). Then write the optimality system (12) as a generalized equation of the form

$$\text{find } \xi \in X \text{ such that } f(\xi) + \mathcal{H}(\xi) \ni 0,$$

where

$$\xi \mapsto f(\xi) = \begin{pmatrix} \dot{x} - g(x, u) \\ \dot{\psi} + \nabla_x H(x, u, \psi) \\ \nabla_u H(x, u, \psi) \end{pmatrix} \quad \text{and} \quad \xi \mapsto \mathcal{H}(\xi) = \begin{pmatrix} 0 \\ 0 \\ \mathcal{V}(\xi) \end{pmatrix}$$

for

$$\mathcal{V}(\xi) = \{v \in L^\infty \mid v(t) \in N_U(u(t)) \text{ for a.e. } t \in [0,1]\}.$$

Next, we introduce a discrete approximation of the optimality system (12). Let N be a natural number, let $h = 1/N$ be the mesh spacing, and let $t_i = ih$, $i \in \{0, 1, \ldots, N\}$. Denote by $PL_0^N(\mathbb{R}^n, [0,1])$ the space of piecewise linear and continuous functions x over the grid $\{t_i\}$ with values in $\mathbb{R}^n$ and such that $x(0) = 0$, by $PL_1^N(\mathbb{R}^n, [0,1])$ the space of piecewise linear and continuous functions ψ over the grid $\{t_i\}$ with values in $\mathbb{R}^n$ and such that $\psi(1) = 0$, and by $PC^N(\mathbb{R}^m, [0,1])$ the space of piecewise constant and continuous from the right functions u over the grid $\{t_i\}$ with values in $\mathbb{R}^m$. Clearly, $PL_0^N(\mathbb{R}^n, [0,1])$ and $PL_1^N(\mathbb{R}^n, [0,1])$ are subsets of $W^{1,\infty}(\mathbb{R}^n, [0,1])$ and $PC^N(\mathbb{R}^m, [0,1]) \subset L^\infty(\mathbb{R}^m, [0,1])$. Then introduce the products $X^N = PL_0^N(\mathbb{R}^n, [0,1]) \times PL_1^N(\mathbb{R}^n, [0,1]) \times PC^N(\mathbb{R}^m, [0,1])$ as an approximation space for the triple (x, ψ, u). We identify $x \in PL_0^N(\mathbb{R}^n, [0,1])$ with the vector $(x^0, \ldots, x^N)$ of its values at the mesh points (and similarly for ψ), and $u \in PC^N(\mathbb{R}^m, [0,1])$ with the vector $(u^0, \ldots, u^{N-1})$ of the values of u in the mesh subintervals.

Now suppose that, for a fixed natural N, a function $\tilde{\xi}^N = (\tilde{x}^N, \tilde{u}^N, \tilde{\psi}^N) \in X^N$ is found that satisfies the following discrete-time system obtained as a discrete approximation of (12):

$$\begin{cases} \dot{x}^i = g(x^i, u^i), \quad x^0 = 0, \\ \dot{\psi}^i = \nabla_x H(x^i, u^i, \psi^{i+1}), \quad \psi^N = 0, \qquad i = 0, 1, \ldots, N-1, \\ 0 \in \nabla_u H(x^i, u^i, \psi^i) + N_U(u^i), \end{cases} \tag{13}$$

where, since x and ψ are piecewise linear, $\dot{x}^i = (x^{i+1} - x^i)/h$ and similarly for ψ. Clearly, system (13) is a first-order necessary condition of a discrete-time optimal control problem representing a discrete approximation of problem (10)–(11).

Let $\bar{\xi} = (\bar{x}, \bar{u}, \bar{\psi})$ be a solution of the optimality system (12) and suppose that the mapping $f + \mathcal{H}$ is strongly subregular at $\bar{\xi}$ for 0. Then, from the definition of strong subregularity displayed in Lecture 11, we obtain that there exist positive scalars r and κ such that if $\tilde{\xi}^N \in I\!B_r(\bar{\xi})$ is a solution of the discrete-time optimality system (13), then

$$\|\tilde{\xi}^N - \bar{\xi}\| \leq \kappa d(0, f(\tilde{\xi}^N) + \mathcal{H}(\tilde{\xi}^N)).$$

Let $\tilde{z}^N$ be defined as

$$\tilde{z}^N(t) = \begin{pmatrix} g(\tilde{x}^N(t_i), \tilde{u}^N(t_i)) - g(\tilde{x}^N(t), \tilde{u}^N(t)) \\ \nabla_x H(\tilde{x}^N(t_i), \tilde{u}^N(t_i), \tilde{\psi}^N(t_{i+1})) - \nabla_x H(\tilde{x}^N(t), \tilde{u}^N(t), \tilde{\psi}^N(t)) \\ \nabla_u H(\tilde{x}^N(t_i), \tilde{u}^N(t_i), \tilde{\psi}^N(t_i)) - \nabla_u H(\tilde{x}^N(t), \tilde{u}^N(t), \tilde{\psi}^N(t)) \end{pmatrix}$$

for $t \in [t_i, t_{i+1})$. Then $\tilde{z}^N \in f(\tilde{\xi}^N) + \mathcal{H}(\tilde{\xi}^N)$ and therefore

$$\|\tilde{\xi}^N - \bar{\xi}\| \leq \kappa \|\tilde{z}^N\|.$$

Hence,

$$\begin{aligned} \|\tilde{\xi}^N - \bar{\xi}\| \leq \kappa \max\nolimits_{0 \leq i \leq N-1} \sup\nolimits_{t_i \leq t \leq t_{i+1}} \big[&\|g(\tilde{x}^N(t_i), \tilde{u}^N(t_i)) - g(\tilde{x}^N(t), \tilde{u}^N(t_i))\| \\ &+ \|\nabla_x H(\tilde{x}^N(t_i), \tilde{u}^N(t_i), \tilde{\psi}^N(t_{i+1})) - \nabla_x H(\tilde{x}^N(t), \tilde{u}^N(t_i), \tilde{\psi}^N(t))\| \\ &+ \|\nabla_u H(\tilde{x}^N(t_i), \tilde{u}^N(t_i), \tilde{\psi}^N(t_i)) - \nabla_u H(\tilde{x}^N(t), \tilde{u}^N(t_i), \tilde{\psi}^N(t))\| \big]. \end{aligned}$$

Observe that here $\tilde{x}^N$ is a piecewise linear across the grid $\{t_i\}$ with uniformly bounded derivative, since both $\tilde{x}^N$ and $\tilde{u}^N$ are in some L_∞ neighborhood of $\bar{x}$ and $\bar{u}$, respectively. Taking into account that the functions $g = \nabla_\psi H$, $\nabla_x H$, and $\nabla_u H$ are continuously differentiable, we obtain that the norm of the residual is bounded by a constant independent of N times the mesh size h. Thus, we come to the following result:

Theorem 18.5. *Assume that the mapping $f + \mathcal{H}$ of the optimality system* (12) *is strongly subregular at $\bar{\xi} = (\bar{x}, \bar{u}, \bar{\psi})$ for* 0. *Then there exist constants b and c such that for every $N = 1, \ldots,$ if $\xi^N = (x^N, u^N, \psi^N)$ is a solution to the discretized system* (13) *contained in $I\!B_b(\bar{\xi})$, then*

$$\|x^N - \bar{x}\|_{W^{1,\infty}} + \|u^N - \bar{u}\|_{L^\infty} + \|\psi^N - \bar{\psi}\|_{W^{1,\infty}} \leq ch.$$

Proof of Theorem 18.2.

We will first show that the triple $(\bar{x}^N, \bar{\psi}^N, \bar{u}^N)$ being a solution of (6) is uniformly bounded by a constant independent of N. Choose an arbitrary $\tilde{u}$ from the nonempty set U and let $\tilde{u}_i^N = \tilde{u}$ for all $i = 0, 1, \ldots, N-1$. Let $\tilde{x}^N$ solve the state equation in (6) for $u = \tilde{u}^N$. Then, by the optimality of $\bar{u}^N$, utilizing the assumed positive definiteness of R and the positive semidefiniteness of Q, we obtain

$$\mu \sum_{i=0}^{N-1} h\|\bar{u}_i^N\|^2 \le \sum_{i=0}^{N-1} h(\bar{u}_i^N)^{\mathsf{T}} R\bar{u}_i^N \le J^N(\tilde{x}_i^N, \tilde{u}_i^N),$$

where μ is the constant in the coercivity condition (4) in Lecture 16. Taking norms in $J^N(\tilde{x}_i^N, \tilde{u}_i^N)$ and denoting by $\|\cdot\|_2$ the L^2 norm, we have

$$\mu \sum_{i=0}^{N-1} h\|\bar{u}_i^N\|^2 \le \sum_{i=0}^{N-1} h\|Q\|\|\tilde{x}_i^N\|^2 + \|R\|\|\tilde{u}\|_2^2. \tag{14}$$

We prove next uniform boundedness of $\tilde{x}^N$ by using the discrete Gronwall inequality.

The discretized state equation for $\tilde{x}^N$ yields

$$\tilde{x}_{i+1}^N = \sum_{j=0}^{i} h(A\tilde{x}_j^N + B\tilde{u}_j^N), \quad \tilde{x}_0^N = x^0. \tag{15}$$

Therefore

$$\|\tilde{x}_{i+1}^N\| \le h\|A\| \sum_{j=0}^{i} \|\tilde{x}_j^N\| + \|B\|\|\tilde{u}\|_2,$$

for $c \ge \max\{\|A\|, \|B\|\}$; we have

$$\|\tilde{x}_{i+1}^N\| \le ch \sum_{j=0}^{i} \|\tilde{x}_j^N\| + c\|\tilde{u}\|_2.$$

Applying the first part of the discrete Gronwall inequality, we obtain

$$\|\tilde{x}_{i+1}^N\| \le c\|\tilde{u}\|_2(1 + ch)^i \le c\|\tilde{u}\|_2(1 + ch)^N \quad i = 0, 1, \ldots, N-1.$$

Since $(1 + ch)^N \le e^c$, we conclude that the sequence $\tilde{x}_i^N$ is uniformly bounded, e.g., by a constant c_1:

$$\max_{1 \le i \le N} \|\tilde{x}_i^N\| \le c_1,$$

and then, using (14), there exists another positive constant c_2 such that

$$\|\bar{u}^N\| = \sum_{j=0}^{N-1} h\|\bar{u}_j^N\|^2 \le c_2. \tag{16}$$

Note that in the preceding lines, as well as in the rest of the proof, c_j, $j = 1, \ldots,$ denotes a generic constant that may be different in different relations but depends only on the matrices A, B, Q, R and the set U.

We will now show the uniform boundedness of the optimal state, control, and co-state variables satisfying (4). By taking norms in (14), we obtain

$$\|\bar{x}_{i+1}^N\| \leq \sum_{j=0}^{i} (h\|A\|\|\bar{x}_j^N\| + h\|B\|\|\bar{u}_j^N\|). \tag{17}$$

The Cauchy–Schwarz inequality

$$\left(\sum_{i=1}^{m} \|v_i w_i\|\right)^2 \leq \sum_{i=1}^{m} \|v_i\|^2 \sum_{i=1}^{m} \|w_i\|^2$$

applied to the second term in the right side of (17) gives us

$$\begin{aligned}\|\bar{x}_{i+1}^N\| &\leq \sum_{j=0}^{i} (h\|A\|\|\bar{x}_j^N\| + h\|B\|\|\bar{u}_j^N\|) \\ &\leq \sum_{j=0}^{i} h\|A\|\|\bar{x}_j^N\| + \|B\| \left(\sum_{i=0}^{N-1} h\|\bar{u}_i^N\|^2\right)^{1/2},\end{aligned}$$

where we take into account that $\sum_{i=0}^{N-1} h = 1$. Then the discrete Gronwall inequality combined with (16) results in

$$\|\bar{x}_i^N\| \leq c_3 \quad \text{for all } i = 0, 1, \ldots, N. \tag{18}$$

Taking norms in the co-state equation in (4) gives us

$$\|\bar{\psi}_i^N\| \leq \sum_{j=i+1}^{N} (h\|A\|\|\bar{\psi}_j^N\| + h\|Q\|\|\bar{x}_j^N\|).$$

Applying the second part (backward) of the discrete Gronwall inequality and using (18), we get

$$\|\bar{\psi}_i^N\| \leq c_4 \quad \text{for all } i = 0, 1, \ldots, N. \tag{19}$$

Consider now the variational inequality in (4). For the optimal $\bar{u}_i^N$ we have

$$0 \leq (R\bar{u}_i^N + B^\top \bar{\psi}_i^N)^\top (\tilde{u} - \bar{u}_i^N),$$

which gives us

$$(\bar{u}_i^N)^\top R\bar{u}_i^N \leq \tilde{u}^\top R\bar{u}_i^N + (\bar{\psi}_i^N)^\top B\tilde{u} - (\bar{\psi}_i^N)^\top B\bar{u}_i^N.$$

Using the positive definiteness of R and taking norms, we obtain

$$\mu\|\bar{u}_i^N\|^2 \leq (\|R\|\|\tilde{u}\|_2 + \|B\|\|\bar{\psi}_i^N\|)\|\bar{u}_i^N\| + \|B\|\|\bar{\psi}_i^N\|\|\tilde{u}\|_2.$$

Hence, (19) yields

$$\|\bar{u}_i^N\|^2 \leq c_5\|\bar{u}_i^N\| + c_6.$$

This implies that the sequence of optimal discrete controls $\bar{u}^N = (\bar{u}_0^N, \ldots, \bar{u}_{N-1}^N)$ is uniformly bounded. Indeed, taking maximum with respect to i on both sides of the last inequality, and denoting $a_N = \max_{0\leq i\leq N-1}\|\bar{u}_i^N\|$ results in

$$a_N^2 \leq c_5 a_N + c_6.$$

By solving this quadratic inequality, we obtain

$$\max_{0\leq i\leq N-1}\|\bar{u}_i^N\| \leq c_7, \tag{20}$$

where the constant c_7 depends only on the matrices A, B, Q, and R and the set U.

Let us now proceed with the proof of the estimate in the statement of the theorem. For $t \in [t_i, t_{i+1})$, $i = 0, 1, \ldots, N-1$, let

$$\begin{aligned}
p^N(t) &= A(\bar{x}^N(t_i) - \bar{x}^N(t)),\\
s^N(t) &= A^\mathsf{T}(\bar{\psi}^N(t_{i+1}) - \bar{\psi}^N(t)) + Q(\bar{x}^N(t_{i+1}) - \bar{x}^N(t)),\\
r^N(t) &= -B^\mathsf{T}(\bar{\psi}^N(t_i) - \bar{\psi}^N(t)).
\end{aligned}$$

By virtue of the control $\bar{u}^N$ being piecewise constant and

$$\dot{\bar{x}}^N(t) = \frac{\bar{x}^N(t_{i+1}) - \bar{x}^N(t_i)}{h}, \quad \dot{\bar{\psi}}^N(t) = \frac{\bar{\psi}^N(t_{i+1}) - \bar{\psi}^N(t_i)}{h} \quad \text{for } t \in [t_i, t_{i+1}),$$

for $i = 0, 1, \ldots, N-1$, the discrete optimality system (6) can be written as follows: for all $t \in [t_i, t_{i+1})$, $i = 0, 1, \ldots, N-1$, and $t = 1$,

$$\begin{cases}
\dot{\bar{x}}^N(t) = A\bar{x}^N(t) + B\bar{u}^N(t) + p^N(t), & \bar{x}^N(0) = x^0,\\
\dot{\bar{\psi}}^N(t) = -A^\mathsf{T}\bar{\psi}^N(t) - Q\bar{x}^N(t) - s^N(t), & \bar{\psi}^N(1) = 0,\\
R\bar{u}^N(t) + B^\mathsf{T}\bar{\psi}^N(t) + N_U(\bar{u}^N(t)) \ni r^N(t).
\end{cases} \tag{21}$$

Observe that this system has the same form as system (9) in the preceding Lecture 17 with a particular choice of the parameters. Specifically, $(\bar{x}^N, \bar{u}^N, \bar{\psi}^N)$ is the solution of that system for the parameter value $z^N := (p^N, s^N, r^N)$, while $(\bar{x}, \bar{u}, \bar{\psi})$ is the solution for $z = (p, s, r) = (0, 0, 0)$, that is, of the optimality system (3). Then, from Theorem 17.2, the solution mapping of system (21) is a globally Lipschitz continuous function acting from L^∞ to L^∞. The state and adjoint trajectories are continuous functions in time; hence, their L^∞ norms are the same as the uniform C norms. Summarizing there exists a constant c_8 such that

$$\|\bar{u}^N - \bar{u}\|_{L^\infty} + \|\bar{x}^N - \bar{x}\|_C + \|\bar{\psi}^N - \bar{\psi}\|_C \leq c_8\|z^N\|_{L^\infty}.$$

To finish the proof, we need to show that

$$\|z^N\|_{L^\infty} = \max\{\|p^N\|_{L^\infty}, \|s^N\|_{L^\infty}, \|r^N\|_{L^\infty}\} \leq c_9 h \tag{22}$$

for some constant c_9 and all N (remember that $h = 1/N$). First, observe that $\bar{x}^N$ is piecewise linear across the grid $\{t_i\}$, and then

$$\|\bar{x}^N(t) - \bar{x}^N(t_i)\| \leq \|\bar{x}^N(t_{i+1}) - \bar{x}^N(t_i)\| \quad \text{for } t \in [t_i, t_{i+1}],\ i = 0, 1, \ldots, N-1.$$

Then, from the definition of p^N, for $t \in [t_i, t_{i+1}]$ and $i = 0, 1, \ldots, N-1$, we have

$$\|p^N(t)\| \leq \|\bar{x}^N(t_{i+1}) - \bar{x}^N(t_i)\| \leq h\|A\|\|\bar{x}^N(t_i)\| + h\|B\|\|\bar{u}_i^N\|.$$

Taking norms in the state equation in (5) and using (18) and (20), this last estimate gives us

$$\|p^N\|_{L^\infty} \leq c_{10} h.$$

By repeating this argument for the discrete adjoint equation in (6), but now applying the second (backward) part of the discrete Gronwall inequality, we get the same order of magnitude for s^N and then for r^N. This proves (22). The proof is complete. □

Lecture 19
Optimal Feedback Control

This lecture introduces to a classical topic of control theory: design of a feedback that, by observing the state at the current instant of time, produces automatically an optimal control, which is then applied to the control system. In order to keep the presentation simple, we focus on the linear-quadratic optimal control problem from Lecture 16:

$$\text{minimize} \quad J(x,u) = \frac{1}{2}\int_0^1 x(t)^{\mathsf{T}} Q x(t) + u(t)^{\mathsf{T}} R u(t) dt \tag{1}$$

subject to

$$\begin{array}{l} \dot{x}(t) = Ax(t) + Bu(t) \quad \text{for a.e. } t \in [0,1], \qquad x(0) = x^0, \\ u(t) \in U \quad \text{for a.e. } t \in [0,1], \\ x \in W^{1,2}(I\!R^n, [0,1]),\ u \in L^2(I\!R^m, [0,1]), \end{array} \tag{2}$$

under the standing assumptions that the matrix R is positive definite, while Q is positive semidefinite, both R and Q are symmetric, and U is a nonempty, closed, and convex subset of $I\!R^m$. Then, problem (1)–(2) has a unique solution $(\bar{x}, \bar{u})$ and there exists an optimal adjoint trajectory $\bar{\psi}$ such that $(\bar{x}, \bar{\psi}, \bar{u})$ is the unique solution of the following system:

$$\begin{array}{ll} \dot{x}(t) = Ax(t) + Bu(t), & x(0) = x^0, \\ \dot{\psi}(t) = -A^{\mathsf{T}}\psi(t) - Qx(t), & \psi(1) = 0, \\ 0 \in Ru(t) + B^{\mathsf{T}}\psi(t) + N_U(u(t)). & \end{array} \tag{3}$$

Conversely, if $(\bar{x}, \bar{\psi}, \bar{u})$ solves (3), then $(\bar{x}, \bar{u})$ is the optimal solution of problem (1)–(2). From Lecture 16 we also know that there exists a representative (selection) of the optimal control being an L^2 function in general, which is Lipschitz continuous in $t \in [0,1]$ and the system (3) holds for all $t \in [0,1]$. Recall that the positive definiteness of R is equivalent to the existence of a constant $\mu > 0$ such that

$$u^{\mathsf{T}} R u \geq \mu \|u\|^2 \quad \text{for all } u \in I\!R^m. \tag{4}$$

© The Author(s), under exclusive license to Springer Nature Switzerland AG 2021

A. L. Dontchev, *Lectures on Variational Analysis*, Applied Mathematical Sciences 205, https://doi.org/10.1007/978-3-030-79911-3_19

Let us first consider the case where there are no constraints on the control; that is, $U = I\!R^m$. Then, inasmuch as $N_{I\!R^m}(u) = \{0\}$ for any u, the variational inequality in (3) becomes an equation that, since the matrix R is nonsingular, provides the following expression for the optimal control:

$$u(t) = -R^{-1}B^{\mathsf{T}}\psi(t) \quad \text{for all } t \in [0,1]. \tag{5}$$

Let the matrix-valued function K be the (unique) solution of the following Riccati-type matrix differential equation:

$$\dot{K} = -Q - A^{\mathsf{T}}K - KA + KBR^{-1}B^{\mathsf{T}}K, \quad K(1) = 0.$$

A direct calculation shows that when

$$\psi(t) = K(t)x(t), \tag{6}$$

the control u is determined by (5), and x is a solution of the state equation in (2) for that u, then the triple (x, ψ, u) satisfies the optimality system (3) for any initial condition x^0. Hence, this is the optimal triple $(\bar{x}, \bar{\psi}, \bar{u})$. Substituting (6) into (5), we obtain the optimal control in a *feedback* form:

$$u(t) = -R^{-1}B^{\mathsf{T}}K(t)x(t) \quad \text{for all } t \in [0,1]. \tag{7}$$

In technical terms, this means that based on (7), it is possible to construct a device, a *feedback controller*, which measures the current state $x(t)$ and generates a control $u(t)$ that is applied to the system at time t producing the optimal state $x(t)$ (Fig. 19.1).

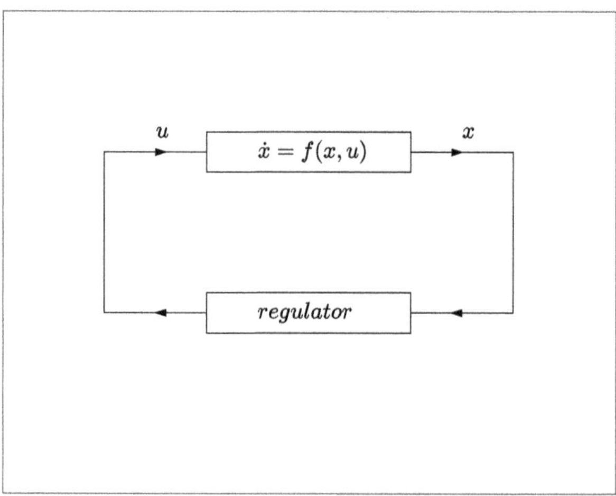

Fig. 19.1: Feedback control

In this lecture we will establish the *existence* of an optimal feedback control for the control-constrained problem (1)–(2). The question of how to find an explicit form of such an optimal feedback remains open. In the following lecture we will show an approach to find an approximation of the optimal feedback control.

Let $b > 0$ be fixed and let $\tau \in [0, 1]$. For any $y \in I\!B_b(x^0)$ consider problem (1)–(2) on the time interval $[\tau, 1]$ with initial state $x(\tau) = y$. The associated optimality system has the form

$$\begin{cases} \dot{x}(t) = Ax(t) + Bu(t), \quad x(\tau) = y, \\ \dot{\psi}(t) = -A^{\mathsf{T}}\psi(t) - Qx(t), \quad \psi(1) = 0, \\ 0 \in Ru(t) + B^{\mathsf{T}}\psi(t) + N_U(u(t)) \;\text{ for all }\; t \in [\tau, 1]. \end{cases} \tag{8}$$

Let $\bar{u}[\tau, y]$ be the optimal control for that problem and let $\bar{x}[\tau, y]$ and $\bar{\psi}[\tau, y]$ be the associated optimal state and co-state trajectories. We begin with a lemma the proof of which is given at the end of the lecture. As in the preceding lecture, the matrices A, B, R, Q, the initial condition x^0, and the set U are said to be the data of problem (1)–(2).

Lemma 19.1. *For any $b > 0$ there exist positive constants c, l and κ such that every $\tau \in [0, 1]$ and $y \in I\!B_b(x^0)$ and the optimal control $\bar{u}[\tau, y]$ satisfies:*

$$\|\bar{u}[\tau, y](t)\| + \|\bar{x}[\tau, y](t)\| + \|\bar{\psi}[\tau, y](t)\| \le c \;\textit{ for every }\; t \in [\tau, 1].$$

$$\|\bar{u}[\tau, y](t) - \bar{u}[\tau, y](t')\| \le l|t - t'| \;\textit{ for every }\; t, t' \in [\tau, 1],$$

and

$$\|\bar{u}[\tau, y](t) - \bar{u}[\tau, y'](t)\| \le \kappa\|y - y'\| \;\textit{ for every }\; y, y' \in I\!B_b(x^0) \;\textit{ and every }\; t \in [\tau, 1].$$

Moreover, the constants c, l, and κ depend only on the constant b and the data of problem (1)–(2).

Next, we give a definition of what we mean by an optimal feedback control for problem (1)–(2):

Optimal Feedback Control. *A function $u^* : [0, 1] \times I\!R^n \to U$ is said to be an optimal feedback control for problem* (1)–(2) *if for every $(\tau, y) \in [0, 1] \times I\!R^n$ the equation*

$$\dot{x}(t) = Ax(t) + Bu^*(t, x(t)), \quad x(\tau) = y \tag{9}$$

has a unique solution $x^*[\tau, y]$ on $[\tau, 1]$, and, for

$$u^*[\tau, y](t) := u^*(t, x^*[\tau, y](t)),$$

the reduced to $[\tau, 1]$ objective function

$$J_\tau(x, u) = \frac{1}{2}\int_\tau^1 x(t)^{\mathsf{T}}Qx(t) + u(t)^{\mathsf{T}}Ru(t)dt$$

satisfies

$$J_\tau(x^*[\tau, y], u^*[\tau, y]) \le J_\tau(x, u)$$

for any feasible control u on the interval $[\tau, 1]$ and a corresponding state trajectory x obtained by solving the state equation (9) *on* $[\tau, 1]$ *for u and the initial condition* $x(\tau) = y$.

Note that the finiteness of the optimal value combined with condition (4) implies that for each $\tau \in [0, 1]$ we have $u^*[\tau, y] \in L^2(I\!R^m, [\tau, 1])$. The main result of this lecture follows:

Theorem 19.2. *There exists an optimal feedback control for problem* (1)–(2) *that is a locally Lipschitz continuous function.*

The proof of this theorem employs the following fundamental principle in optimization adjusted to the problem considered:

Dynamic Programming Principle. *For $\tau \in [0, 1]$ and $y \in I\!R^n$, let $\bar{u}[\tau, y]$ be the optimal control for problem* (1)–(2) *on the interval $[\tau, 1]$ with the initial condition $x(\tau) = y$ and let $\bar{x}[\tau, y]$ be the associated optimal state trajectory. Then for any* $\theta \in (\tau, 1)$

$$\bar{u}[\theta, \bar{x}[\tau, y](\theta)](t) = \bar{u}[\tau, y](t) \quad \textit{for all} \;\; t \in [\theta, 1].$$

The dynamic programming principle has a simple intuitive justification. If for some $\theta \in [\tau, 1]$ the optimal control $\bar{u}[\theta, \bar{x}[\tau, y](\theta)]$ was different from $\bar{u}[\tau, y]$ restricted to $[\theta, 1]$, then we would be able to reduce the cost for $\bar{u}[\tau, y]$ when we switch at moment θ to the control $\bar{u}[\theta, \bar{x}[\tau, y](\theta)]$; this contradicts the optimality of $\bar{u}[\tau, y]$. For example, if the fastest route from New York to Los Angeles passes through Denver, then the portion of the route from Denver to Los Angeles is the fastest route from Denver to Los Angeles.

Proof of Theorem 19.2. By the dynamic programming principle, for any $y \in I\!R^n$, $\tau \in [0, 1)$, $\tau' \in (\tau, 1]$, and $t \in [\tau', 1)$,

$$\bar{u}[\tau, y](t) = \bar{u}[\tau', \bar{x}[\tau, y](\tau')](t).$$

Then from Lemma 19.1 we obtain that there exists a constant c_1 such that

$$\begin{aligned} \|\bar{u}[\tau, y](t) - \bar{u}[\tau', y](t)\| &= \|\bar{u}[\tau', \bar{x}[\tau, y](\tau')](t) - \bar{u}[\tau', y](t)\| \\ &\le \kappa \|\bar{x}[\tau, y](\tau') - y\| \le c_1 |\tau' - \tau|, \end{aligned} \tag{10}$$

where we use the equalities

$$\bar{x}[\tau, y](\tau') - y = \int_\tau^{\tau'} \dot{\bar{x}}[\tau, y](s) ds = \int_\tau^{\tau'} A\bar{x}[\tau, y](s) + B\bar{u}[\tau, y](s) ds$$

and the uniform boundedness of $\bar{x}[\tau, y]$ and $\bar{u}[\tau, y]$ established in Lemma 19.1. Observe that the constant c_1 depends only on b, the matrices in problem (1)–(2), and the set U.

For $t \in [0, 1]$ and $x \in \mathbb{R}^n$, we define the optimal feedback control as

$$u^*(t, x) = \bar{u}[t, x](t).$$

Using the inequalities established in Lemma 19.1 and (10), we have

$$\begin{aligned}\|u^*(t, x) - u^*(t', x')\| &= \|\bar{u}[t, x](t) - \bar{u}[t', x'](t')\| \\ &\le \|\bar{u}[t, x](t) - \bar{u}[t', x](t)\| + \|\bar{u}[t', x](t) - \bar{u}[t', x'](t)\| \\ &\quad + \|\bar{u}[t', x'](t) - \bar{u}[t', x'](t')\| \\ &\le \kappa\|x - x'\| + (c_1 + l)|t - t'|.\end{aligned}$$

It is well known that an ordinary differential equation with a locally Lipschitz continuous right side has a unique solution. The remaining condition for u^* follows from the optimality of $\bar{u}$. □

Proof of Lemma 19.1.

In order to avoid the use of too many constants, as in the proof of Theorem 18.2, in further lines we denote by c a generic constant, which depends only on the constant b and the data of the problem, that is, the matrices A, B, Q, and R, the initial point x^0, and the set U, and which may be different in different relations.

Choose any $\tau \in [0, 1)$ and $y \in \mathbb{B}_b(x^0)$. Recall the Cauchy formula for the state equation on the interval $[\tau, 1]$ with the initial condition $x(\tau) = y$:

$$x(t) = e^{A(t-\tau)}y + \int_\tau^t e^{A(t-s)}Bu(s)ds. \tag{11}$$

Let $\tilde{v} \in U$, let $\tilde{u}(t) = \tilde{v}$ for all $t \in [0, 1]$, and let $\tilde{x}$ solve the state equation for $u = \tilde{u}$. Then, by taking norms in the Cauchy formula for $\tilde{x}$ resulting from $\tilde{u}$, we get

$$\|\tilde{x}(t)\| \le c(\|x(0)\| + b) + c\|\tilde{v}\| \le c \text{ for all } t \in [\tau, 1].$$

By the definiteness assumptions for R and Q and by optimality,

$$\mu\|\bar{u}[\tau, y]\|^2_{L^2(\mathbb{R}^m,[\tau,1])} \le J_\tau(\bar{x}[\tau, y], \bar{u}[\tau, y]) \le J(\bar{x}[\tau, y], \bar{u}[\tau, y]) \le J(\tilde{x}, \tilde{u}) \le c;$$

hence,

$$\|\bar{u}[\tau, y]\|_{L^2(\mathbb{R}^m,[\tau,1])} \le c. \tag{12}$$

From (11), for any $t \in [\tau, 1]$ we have

$$\begin{aligned}\|\bar{x}[\tau, y](t)\| &\le e^{\|A\|t}\|y\| + \int_\tau^t \|e^{A(t-s)}B\bar{u}[\tau, y](s)\|ds \\ &\le c(\|x(0) + b) + c\int_\tau^1 \|\bar{u}[\tau, y](s)\|ds \le cb + c\|\bar{u}[\tau, y]\|_{L^2(\mathbb{R}^m,[\tau,1])},\end{aligned}$$

where we obtain the last inequality by using the Cauchy–Schwarz inequality. Then, utilizing (12), we conclude that

$$\max_{t\in[\tau,1]} \|\bar{x}[\tau, y](t)\| \le c. \tag{13}$$

By applying the (backward) Cauchy formula

$$\bar{\psi}[\tau, a](t) = \int_t^1 e^{-A^\top(t-s)} Q\bar{x}[\tau, y](s)ds \tag{14}$$

for the adjoint equation in (8), and using (13), we obtain

$$\max_{t\in[\tau,1]} \|\bar{\psi}[\tau, y](t)\| \le c. \tag{15}$$

From the variational inequality in (8) we have

$$(R\bar{u}[\tau, y](t) + B^\top\bar{\psi}[\tau, y](t))^\top(\tilde{v} - \bar{u}[\tau, y](t)) \ge 0 \text{ for every } t \in [\tau, 1].$$

Hence,

$$\begin{aligned}\mu\|\bar{u}[\tau, y](t)\|^2 &\le \bar{u}[\tau, y](t)^\top R\bar{u}[\tau, y](t)\\ &\le |\bar{u}[\tau, y](t)^\top(R\tilde{v} - B^\top\bar{\psi}[\tau, y](t)) + \tilde{v}^\top B^\top\bar{\psi}[\tau, y](t)|,\end{aligned}$$

which, because of (15), yields

$$\max_{t\in[\tau,1]} \|\bar{u}[\tau, y]\| \le c.$$

Together with (13) and (14), this establishes the first inequality in the statement of the lemma.

Proceeding with the proof, let $t, t' \in [\tau, 1]$. Utilizing the variational inequality in (8), we have

$$(R\bar{u}[\tau, y](t) + B^\top\bar{\psi}[\tau, y](t))^\top(\bar{u}[\tau, y](t') - \bar{u}[\tau, y](t)) \ge 0$$

and

$$(R\bar{u}[\tau, y](t') + B^\top\bar{\psi}[\tau, y](t'))^\top(\bar{u}[\tau, y](t) - \bar{u}[\tau, y](t')) \ge 0.$$

Adding these two inequalities results in

$$\begin{aligned}&\mu\|(\bar{u}[\tau, y](t') - \bar{u}[\tau, y](t)\|^2\\ &\quad\le (\bar{u}[\tau, y](t') - \bar{u}[\tau, y](t))^\top R(\bar{u}[\tau, y](t') - \bar{u}[\tau, y](t))\\ &\quad\le B^\top(\bar{\psi}[\tau, y](t) - \bar{\psi}[\tau, y](t'))^\top(\bar{u}[\tau, y](t') - \bar{u}[\tau, y](t)).\end{aligned} \tag{16}$$

By using the Cauchy formula (14) for the adjoint equation, (13) and (15), we obtain that the function $t \mapsto \bar{\psi}[\tau, y](t)$ is Lipschitz continuous in $t \in [\tau, 1]$ with a Lipschitz constant that depends only on b and the data of the problem. Using this in (16) gives us the second inequality in the statement of the lemma for some constant $l > 0$.

Make the change of variables $z = x - y$ in (8) obtaining the system

$$\begin{cases} \dot{z}(t) = Az(t) + Bu(t) + Ay, \quad z(\tau) = 0, \\ \dot{\psi}(t) = -A^{\mathsf{T}}\psi(t) - Qz(t) - Qy, \quad \psi(1) = 0, \\ 0 \in Ru(t) + B^{\mathsf{T}}\psi(t) + N_U(u(t)) \quad \text{for all } t \in [\tau, 1]. \end{cases} \tag{17}$$

In Theorem 17.2 we showed for a system more general than (17) that the optimal control $\bar{u}$ is globally Lipschitz continuous with respect to perturbations in the right side of the optimality system. In further lines we follow the same argument adapted to problem (17) keeping an eye on the dependence of the constants on the data of the problem.

Denote $p = Ay$ and $s = Qy$. Using the mapping T introduced in Lecture 16, the perturbed optimality system (17) is equivalent, because of convexity, to the perturbed linear-quadratic problem

$$\text{minimize} \quad \int_\tau^1 \frac{1}{2}(z(t)^{\mathsf{T}}Qz(t) + u(t)^{\mathsf{T}}Ru(t)) + s^{\mathsf{T}}z(t)dt \tag{18}$$

subject to

$$z = TBu + Tp, \quad u \in \mathcal{U},$$

where the mapping T and the set $\mathcal{U}$ are defined on $[\tau, 1]$ instead of $[0, 1]$. Next, for

$$w(p, s) = B^*T^*s + B^*T^*QTp$$

problem (18) is equivalent to the perturbed variational inequality

$$w(p, s) + \mathcal{A}u + N_{\mathcal{U}}(u) \ni 0. \tag{19}$$

As before, because of (4), the mapping $\mathcal{A}$ satisfies the coercivity condition

$$\langle u, \mathcal{A}u \rangle \geq \mu \|u\|_{L^2}^2 \quad \text{for all } u \in \mathcal{U} - \mathcal{U}.$$

Then, for each $(p, s) = (Ay, Qy)$ problem (18) has a unique solution $\bar{u}[\tau, y]$ and, moreover, the function $y \mapsto \bar{u}[\tau, y]$ is globally Lipschitz continuous from $L^2(\mathbb{R}^n \times \mathbb{R}^n, [\tau, 1])$ to $L^2(\mathbb{R}^m, [\tau, 1])$ with a Lipschitz constant depending only on μ, and hence on R, as well as on the matrices A, B, and Q. Since (p, s) is a constant in time and a linear function of y, the origin space L^2 can be replaced by $\mathbb{R}^n$. For any $y, y' \in \mathbb{R}^n$, let $\bar{y}[\tau, y]$ and $\bar{y}[\tau, y']$ be the solutions of the state equation in (17) for $(\bar{u}[\tau, y], y)$, and $(\bar{u}[\tau, y'], y')$. Taking norms in the difference of the Cauchy formula (11) for the corresponding solutions of the state equation in (17), we obtain that, for some constant c depending only on the matrices A and B,

$$\|\bar{y}[\tau, y](t) - \bar{y}[\tau, y'](t)\| \leq c(\|\bar{u}[\tau, y] - \bar{u}[\tau, y']\|_{L^2} + \|y - y'\|) \quad \text{for all } t \in [\tau, 1].$$

Taking the supremum on the left and using the Lipschitz continuity of the function $y \mapsto \bar{u}[\tau, y]$, we obtain that the optimal trajectory function $y \mapsto \bar{x}[\tau, y]$ is Lipschitz continuous from the $\mathbb{R}^n$ to $C(\mathbb{R}^n, [\tau, 1])$. The same argument for the corresponding adjoint trajectory $\bar{\psi}$ gives us that the function $y \mapsto \bar{\psi}[\tau, y]$ is Lipschitz continuous from $I\!R^n$ to $C(I\!R^n, [\tau, 1])$ as well. Hence, the mapping describing the solution of the optimality system (17) as a function of y is single-valued and globally Lipschitz continuous from the space $I\!R^n$ to the space $C(I\!R^n \times I\!R^n, [\tau, 1]) \times L^2(I\!R^m, [\tau, 1])$.

Now we consider the variational inequality in (17). By repeating the argument before the statement of Theorem 16.1, we obtain that from the equivalence class of optimal control functions $\bar{u}[\tau, a]$, there exists a representative for which the variational inequality holds for all $t \in [\tau, 1]$ and which is Lipschitz continuous with respect to t on $[\tau, 1]$. Moreover, Theorem 19.1 applied to the variational inequality in (17) for $t \in [\tau, 1]$ yields that its solution mapping is Lipschitz continuous with respect to $\psi[\tau, y](t)$. From this we get that the solution mapping $y \mapsto \bar{u}[\tau, y]$ is Lipschitz continuous from $I\!R^n$ to $C(I\!R^m, [\tau, 1])$. As easily seen from the above lines, the corresponding Lipschitz constant depends only on b and the data of the problem. This completes the proof of the lemma. □

Lecture 20
Model Predictive Control

In this lecture we present an algorithmic approximation of the optimal feedback control called *model predictive control* (MPC). In the last several decades MPC has emerged as a new paradigm in control theory with a broad range of applications. We will consider a version of MPC on a finite time interval, which leads to a so-called shrinking horizon formulation. Other popular versions of MPC involve a receding or moving horizon.

We will present a model predictive control algorithm for a perturbed version of the linear-quadratic optimal control problem considered in the last four lectures:

$$\text{minimize} \quad J(x,u) = \frac{1}{2}\int_0^1 x(t)^\mathsf{T} Q x(t) + u(t)^\mathsf{T} R u(t) dt \tag{1}$$

subject to

$$\begin{aligned}
&\dot{x}(t) = Ax(t) + Bu(t) + p(t) \quad \text{for a.e. } t \in [0,1], \qquad x(0) = x^0,\\
&u(t) \in U \quad \text{for a.e. } t \in [0,1],\\
&x \in W^{1,2}(I\!R^n, [0,1]),\ u \in L^2(I\!R^m, [0,1]),
\end{aligned} \tag{2}$$

where now $p(t)$ is a time-varying parameter with a reference value $\bar{p}$ representing uncertainty in the system. When we say "value" of the parameter p, we mean a specific function $t \mapsto p(t)$ representing the evolution of the parameter in time, which is assumed to belong to $L^\infty(I\!R^n, [0,1])$. It is also assumed that some reference value (prediction) $\bar{p}$ is available, which is Lipschitz continuous in time on $[0,1]$. The initial state is fixed at x^0. We adopt the standing assumptions that the matrix R is positive definite, while Q is positive semidefinite, both R and Q are symmetric, and U is a nonempty, closed, and convex subset of $I\!R^m$. On these assumptions, in Lecture 16, we derived a necessary and sufficient optimality condition for problem (1)–(2) without the parameter p, in Lecture 18 we analyzed a discrete approximation of it, and in Lecture 19 we proved the existence of a Lipschitz continuous optimal feedback control. In the sequel $x[u,p]$ denotes the solution of state equation in (2) obtained for control u and parameter p.

© The Author(s), under exclusive license to Springer Nature Switzerland AG 2021

A. L. Dontchev, *Lectures on Variational Analysis*, Applied Mathematical Sciences 205, https://doi.org/10.1007/978-3-030-79911-3_20

The MPC algorithm considered is an iterative procedure employing the discrete approximation of problem (1)–(2) studied in Lecture 18 and based on the Euler scheme over a uniform mesh. Given a natural number N, let $\{t_k\}_0^N$ be a grid on $[0, 1]$ with equally spaced nodes t_k and a step-size $h = 1/N$. To describe the MPC iteration, fix $k \in \{0, 1, \dots, N-1\}$ and assume that a control u^N is already determined on $[0, t_k)$ and applied to the system in (2) with value p of the parameter. Also assume that the resulting value of the state at time t_k, $x[u^N, p](t_k)$, is measured or estimated with an error e_k, that is, the vector

$$x_k^0 := x[u^N, p](t_k) + e_k \tag{3}$$

becomes available at time t_k. Then the control on the time interval $[t_k, t_{k+1})$ is determined by solving the discrete-time optimal control problem

$$\text{minimize} \quad J_k^N(x, u) = \frac{1}{2} \sum_{i=k}^{N-1} h(x_i^{\mathsf{T}} Q x_i + u_i^{\mathsf{T}} R u_i) \tag{4}$$

subject to

$$x_{i+1} = (I + hA)x_i + hBu_i + h\bar{p}_i, \quad x_k = x_k^0, \quad u_i \in U \quad \text{for } i = k, k+1, \dots, N-1. \tag{5}$$

Note that this problem is solved for the reference value $\bar{p}$ of the parameter. For $k = 0$ we take $x_0^0 = x^0 + e_0$. We know that under the standing assumptions, problem (4)–(5) has a unique solution. Denote by $\bar{u}^N = (\bar{u}_k^N, \dots, \bar{u}_{N-1}^N)$ the optimal discrete-time control of this problem. Define the constant in time function

$$u^N(t) = \bar{u}_k^N \quad \text{for } t \in [t_k, t_{k+1}), \tag{6}$$

change k to $k + 1$, and continue the iterations as long as $k < N$ (Fig. 20.1).

When k reaches N, we obtain a piecewise constant function u^N over $[0, 1]$, which we call further the *MPC-generated control*. Note that we keep the final time $t = 1$ fixed, so that the time horizon shrinks at each iteration. As in the preceding two lectures, we identify as data of problem (1)–(2) the matrices A, B, Q, and R, the set U, and the initial condition x^0.

In Lecture 19, for problem (1)–(2) under the standing assumptions but without a parameter p, we established the existence of an optimal feedback control $(t, x) \mapsto u^*(t, x)$, which is locally Lipschitz continuous. A brief inspection of the proof shows that the presence of p does not change this result. When the feedback control u^* is applied to the system in (2) with value p of the parameter and with exact measurement, the resulting function $t \mapsto \hat{u}(t) := u^*(t, x[u^*, p](t))$ is said to be a *realization* of the optimal feedback control. The main result in this lecture is the following theorem, which gives an estimate for the difference between the MPC-generated control and the optimal feedback control:

Theorem 20.1. *For every $\delta > 0$ there exist a constant c such that for every N, for every p with $\|p - \bar{p}\|_{L^\infty} \le \delta$, and for every $e = (e_0, \dots e_{N-1})$ with*

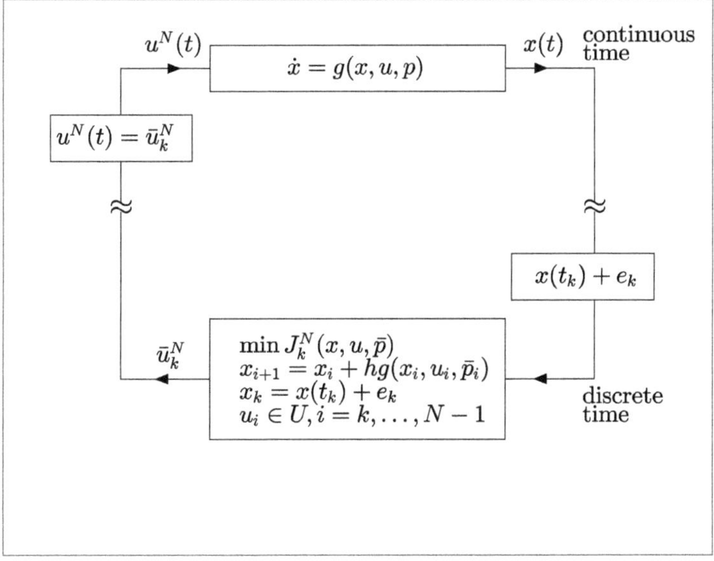

Fig. 20.1: Model predictive control structure

$\max_{k=0,\ldots,N-1} |e_k| \leq \delta$ *the difference between the MPC-generated control* u^N *and the realization* $\hat{u}(t) := u^*(t, x[u^*, p](t))$ *of the optimal feedback control satisfies the estimate*

$$\|u^N(t_i) - \hat{u}(t_i)\| \leq c\Big(h + \|e_i\| + h\sum_{k=0}^{N-1} \|e_k\|\Big), \quad i = 0, \ldots, N-1. \tag{7}$$

This result shows that the MPC-generated control is indeed an approximation of the optimal feedback control, the accuracy of which is determined by the step-size h of the discretization and the measurement error e. Note that the estimate does not depend on the value of the parameter p; the only condition involved is that for (almost) every $t \in [0, 1]$ the values $p(t)$ must belong to a ball with fixed radius around the reference $\bar{p}(t)$. This means that a change of the parameter p affects in the same way (modulo $O(h)$) both the optimal feedback control and the control generated by the MPC algorithm.

The reminder of this lecture is devoted to proving Theorem 20.1. We start with a lemma, the proof of which is presented at the end of the lecture.

Lemma 20.2. *For every* $\delta > 0$ *there exists a positive constant* c *that depends only on the data of the problem and* δ *such that for every* p *with* $\|p - \bar{p}\|_{L^\infty} \leq \delta$*, every* $e = (e_0, \ldots e_{N-1})$ *with* $\max_{k=0,\ldots,N-1} |e_k| \leq \delta$ *and every* $N = 1, 2 \ldots,$

$$\max_{t\in[0,1]} (\|u^N(t)\| + \|x[u^N, p](t)\|) < c.$$

We also use the following standard result:

Continuous Gronwall Inequality. *Let $\alpha \in I\!R$, $\beta : [0,1] \to I\!R_+$ be a continuous function on the interval $[0,1]$ and suppose that a function $w : [0,1] \to I\!R$ satisfies*

$$w(t) \le \alpha + \int_0^t \beta(s)w(s)ds.$$

Then

$$w(t) \le \alpha \exp\left(\int_0^t \beta(s)ds\right) \quad \text{for all } t \in [0,1].$$

Proof of Theorem 20.1. We already mentioned that adding a parameter p to the control system in (2) does not affect the existence of a feedback control u^* claimed in Theorem 19.2. Also observe that due to Lemma 19.1, the optimal state trajectory of the closed-loop system belongs to a ball in $I\!R^n$ whose radius does depend only on the data of the problem and the size of the perturbations p and e. Let $\hat{L}$ be the Lipschitz constant of the optimal feedback control u^* that depends only on these data.

Denote by $\bar{u}^N[k]$ the optimal control for the discrete-time problem (4)–(5) and let $\bar{u}[t_k, y]$ be the optimal control for the continuous-time problem (1)–(2) on the interval $[t_k, 1]$ with the initial state condition $x(t_k) = y$, which is Lipschitz continuous in time t on $[t_k, 1]$. According to Theorem 18.3, there exists a constant c depending only on the data of the problem and the size of p and e (bounded by δ) such that

$$\|\bar{u}_j^N - \bar{u}[k](t_j)\| \le ch. \tag{8}$$

Furthermore, from Lemma 20.2, we know that the values of the MPC-generated control u^N belong to a bounded set in $I\!R^m$ depending only on the data of the problem and the size of p and e. This implies through the Cauchy formula that the mapping $t \mapsto x^N(t) = x[u^N, p](t)$ is Lipschitz continuous on $[0,1]$ and the expression

$$\bar{L} := e^{\|A\|}\|B\| \max_{\tau\in[0,1]} \|u^N(\tau)\|$$

is a Lipschitz constant for that mapping. Denote

$$K = e^{\|A\|}\|B\|\hat{L} \quad \text{and} \quad M = e^{\|A\|}\|B\|(\bar{L}\hat{L} + c),$$

where c is as in (8). For $k = 0, \ldots, N-1$, starting with $d_0 = \|e_0\|$, define recursively the sequence

$$d_{k+1} = e^{Kh}\left(d_k + Mh^2 + Kh\|e_k\|\right).$$

We will now show that for every $k = 1, 2, \ldots, N-1$,

$$\max_{t\in[0,t_k]} \|x^N(t) - \hat{x}(t)\| \le d_k. \tag{9}$$

To do that we use induction in k.

For $k = 0$ we have

$$\|x^N(0) - \hat{x}(0)\| = \|e_0\| = d_0.$$

Pick $0 \le j < N$ and assume that for every $k = 1, 2 \ldots, j$ the estimate (9) holds. We will now prove that it holds for $k = j + 1$. Let $\bar{u}[j, x_j^0]$ be the (unique) optimal control for the continuous-time problem on $[t_j, 1]$ for $\bar{p}$ with the initial condition for the state x_j^0. Also, let $\bar{u}_j^N$ be the optimal control for discrete-time problem for $\bar{p}$ with the initial condition for the state x_j^0. Let $s \in [t_j, t_{j+1})$. Then $u^N(s) = \bar{u}_j^N$ from (6). Utilizing (9), Theorem 19.2 (with the addition of the parameter p that does not change anything), and the definition of $\hat{x}$, we have

$$\begin{aligned}\|u^N(s) - \hat{u}(s)\| &\le \|u^N(s) - \bar{u}[j, x_j^0](t_j)\| + \|\bar{u}[j, x_j^0](t_j) - \hat{u}(s)\| \\ &\le ch + \|u^*(t_j, x_j^0) - u^*(s, \hat{x}(s))\| \\ &\le ch + \hat{L}\left(|t_j - s| + \|x_k^0 - \hat{x}(s)\|\right) \\ &\le ch + \hat{L}\left(h + \|x^N(t_j) - \hat{x}(s)\| + \|e_j\|\right).\end{aligned}$$

Using the Lipschitz continuity of x^N with constant $\bar{L}$, we obtain

$$\|u^N(s) - \hat{u}(s)\| \le ch + \hat{L}(\|x^N(s) - \hat{x}(s)\| + \|e_j\| + \bar{L}h). \tag{10}$$

Invoking the Cauchy formula for the state equation, we get that, for $t \in [t_j, t_{j+1}]$,

$$\begin{aligned}\|x^N(t) - \hat{x}(t)\| &\le \|x^N(t_j) - \hat{x}(t_j)\| + \|\int_{t_j}^t e^{A(t-s)} B(u^N(s)) - \hat{u}(s))ds\| \\ &\le d_j + e^{\|A\|}\|B\| \int_{t_j}^t \|u^N(s) - \hat{u}(s)\|ds \\ &\le d_j + e^{\|A\|}\|B\| \int_{t_k}^t (ch + \hat{L}\|x^N(s) - \hat{x}(s)\| + \hat{L}\|e_j\| + \hat{L}\bar{L}h)ds \\ &\le d_j + \int_{t_k}^t K\|x^N(s) - \hat{x}(s)\|ds + Kh\|e_j\| + Mh^2.\end{aligned}$$

The continuous Gronwall inequality yields

$$\|x^N(t) - \hat{x}(t)\| \le e^{Kh}(d_k + Kh|e_j| + Mh^2) = d_{k+1}.$$

The induction step is complete and so is the proof of (9).

Having (9), it is now easy to obtain (using, e.g., induction again) that

$$d_k = \sum_{i=0}^{k-1} e^{K(k-i)h} L(Mh^2 + Kh\|e_i\|) \quad \text{for all } k = 0, 1 \ldots, N-1;$$

hence,

$$d_k \le c\left(h + h\sum_{i=0}^{N-1} \|e_i\|\right),$$

with a constant c that depends only on data of the problem and δ. This, combined with (9) and (10), yields the desired estimate (7). □

Proof of Lemma 20.2. First, note that u^N defined iteratively in (6) depends on p through the initial state x_k^0 given in (3). The optimality system of problem (4)–(5) has the form

$$\begin{cases} x_{i+1} = (I + hA)x_i + hBu_i + \bar{p}_i, & x_0 = x_k^0, \\ \psi_i = (I + hA^\top)\psi_{i+1} + hQx_{i+1}, & \psi_N = 0, \\ 0 \in Ru_i + B^\top\psi_i + N_U(u_i), \end{cases} \tag{11}$$

for $i = k, k+1, \ldots, N-1$. We use induction with k. According to (6), for $k = 0$ we have $u^N(t) = \bar{u}_0^N$ for $t \in [0, t_1)$, where $\bar{u}^N$ solves (4)–(5) for $k = 0$, $x_0 = x^0 + e_0$, x^0 fixed, and $\|e_0\| \le \delta$. We will prove that

$$\max_{0 \le i \le N-1} \|\bar{u}_i^N\| \le c, \tag{12}$$

where here and further in the proof c is a generic constant that depends only on the data of the problem and δ but not on N and may be different in different relations.

The proof of (12) basically repeats the first part of proof of Theorem 18.2 with some adjustments. In order to make this lecture less dependent on the previous material, the proof will be presented in full with a few shortcuts. Choose an arbitrary $\tilde{u}$ from U and let $\tilde{u}_i^N = \tilde{u}$ for all $i = 0, 1, \ldots, N-1$. Let $\tilde{x}^N$ solve the state equation in (11) for $u = \tilde{u}^N$. Then, by the optimality of $\bar{u}^N$, utilizing the assumed positive definiteness of R and the positive semidefiniteness of Q, we obtain

$$\mu \sum_{i=0}^{N-1} h\|\bar{u}_i^N\|^2 \le \sum_{i=0}^{N-1} h(\bar{u}_i^N)^\top R\bar{u}_i^N \le J^N(\tilde{x}_i^N, \tilde{u}_i^N). \tag{13}$$

The state equation for $\tilde{x}^N$ yields

$$\tilde{x}_{i+1}^N = x^0 + e_0 + \sum_{j=0}^{i} h(A\tilde{x}_j^N + B\tilde{u}_j^N + p_j).$$

Taking norms we obtain

$$\|\tilde{x}_{i+1}^N\| \le \|x^0\| + \delta + ch\sum_{j=0}^{i} \|\tilde{x}_j^N\| + c\|\tilde{u}\|,$$

and then, applying the discrete Gronwall inequality, we obtain

$$\max_{1 \le i \le N} \|\tilde{x}_i^N\| \le c.$$

Hence, from (13), there exists (another) positive constant c such that

$$\sum_{j=0}^{N-1} h\|\bar{u}_j^N\|^2 \leq c. \tag{14}$$

For the optimal discrete-time state, we have

$$\|\bar{x}_{i+1}^N\| \leq \|x^0\| + \delta + \sum_{j=0}^{i}(h\|A\|\|\bar{x}_j^N\| + h\|B\|\|\bar{u}_j^N\|),$$

which gives us

$$\|\bar{x}_{i+1}^N\| \leq \|x^0\| + \delta + \sum_{j=0}^{i} h\|A\|\|\bar{x}_j^N\| + \|B\|\left(\sum_{i=0}^{N-1} h\|\bar{u}_i^N\|^2\right)^{1/2}.$$

The discrete Gronwall inequality combined with (14) yields

$$\max_{1\leq i\leq N} \|\bar{x}_i^N\| \leq c. \tag{15}$$

For the adjoint equation in (11) we get

$$\|\bar{\psi}_i^N\| \leq \sum_{j=i+1}^{N} (h\|A\|\|\bar{\psi}_j^N\| + h\|Q\|\|\bar{x}_j^N\|).$$

Then, using (15), the backward (second) part of the discrete Gronwall inequality gives us

$$\|\bar{\psi}_i^N\| \leq c \quad \text{for all } i = 0, 1, \ldots, N. \tag{16}$$

Consider now the variational inequality in (11). For the optimal $\bar{u}_i^N$ we have

$$0 \leq (R\bar{u}_i^N + B^\top\bar{\psi}_i^N)^\top(\tilde{u} - \bar{u}_i^N),$$

which leads to

$$(\bar{u}_i^N)^\top R\bar{u}_i^N \leq \tilde{u}^\top R\bar{u}_i^N + (\bar{\psi}_i^N)^\top B\tilde{u} - (\bar{\psi}_i^N)^\top B\bar{u}_i^N.$$

Using the positive definiteness of R and taking norms, we obtain

$$\mu\|u_i^N\|^2 \leq (\|R\|\|\tilde{u}\| + \|B\|\|\bar{\psi}_i^N\|)\|u_i^N\| + \|B\|\|\bar{\psi}_i^N\|\|\tilde{u}\|.$$

Hence, by using (16), we get

$$\|u_i^N\|^2 \leq c\|u_i^N\| + c.$$

This quadratic inequality gives us

$$\max_{0 \le i \le N-1} \|\bar{u}_i^N\| \le c,$$

where the constant c could be different from that in the preceding formula, of course. Taking norms in the Cauchy formula for $x[u^N, p]$ leads to (12).

The induction step is completely analogous to the first step, with the only difference that now we consider (11) with $x_k^0 = x[u^N, p](t_k) + e_k$, which turns out to be bounded, by the induction assumption and the boundedness of e_k. □

Bibliographical Remarks and Further Reading

Topics of variational analysis have been covered in books on optimization and control already in the early 1970s, and some of these books are still highly recommended to anybody who wants to go deeper in the area. Among them are the path-breaking monograph on convex analysis by R. T. Rockafellar [70], the lecture notes of I. V. Girsanov [44] on a general theory of optimality conditions, the book of Ioffe and Tichomirov [49] treating the same topic more broadly, and the book by I. Ekeland and R. Temam [43] connecting convex analysis with calculus of variations. A decade later came the book of Clarke [14] on nonsmooth analysis, and after another decade the book of Aubin and Frankowska [4] on set-valued analysis; these works opened new horizons in the area.

In the last 20-plus years several monographs on variational analysis have been published, starting with the foundational treaty by Rockafellar and Wets [73]. Then came the important books by Klatte and Kummer [55] about nonsmooth analysis and applications to optimization, Zalinescu [80] about abstract convex analysis, Borwein and Zhu [10] about variational principles, and the @two-volume book by Mordukhovich [61] about generalized differentiation, with a recent addition [62] by the same author. The book [32] by the author and Rockafellar is devoted to regularity properties of mappings in variational analysis and stability of optimization problems with respect to perturbations. Developments in that direction are covered broadly in the more recent monograph by Ioffe [48]. Methods of variational analysis in control are presented in Clarke et al. [16], see also the more recent monograph [15] by the same author. We should also mention the books by Penot [63] and by Schirotzek [75] on nonsmooth analysis, as well as the book Attouch, Buttazzo and Michaille [1], which is about optimization problems involving PDEs. The reader can find in the bibliographical remarks of these books much more about the history and the development of ideas of variational analysis.

Lecture 1. This lecture presents a brief introduction to optimization focused mainly on concepts and statements used later in the book. Readers who want to learn about fundamentals of optimization beautifully presented may start with the lecture notes

© The Author(s), under exclusive license to Springer Nature Switzerland AG 2021

A. L. Dontchev, *Lectures on Variational Analysis*, Applied Mathematical Sciences 205, https://doi.org/10.1007/978-3-030-79911-3

of Rockafellar [72] and then go deeper with the books by Girsanov [44] and Ioffe and Tikhomirov [49].

Lecture 2. There are various ways to define continuity of set-valued mappings, based on the corresponding definitions of set convergence. In this book we choose the concept of continuity based on the Pompeiu–Hausdorff convergence. Another popular concept of continuity utilizes the weaker Painleve–Kuratowski convergence; more about that could be found in Rockafellar and Wets [73] and Lucchetti [58]. For advanced treatments, see the book Beer [6] and the more recent survey by Royset [74]. Theorem 2.2 is a version of the classical Berge theorem given in Berge [7, Chapter 6, Theorem 3]. Other versions of this result could be found in [34, Chapter 9].

Lecture 3. Theorem 3.1 goes back to Walkup and Wets [79], while Lemma 3.2 is due to Hoffman [47]. Note that in [32] the polyhedral mappings are called "polyhedral convex," while the piecewise polyhedral mappings are called "polyhedral." The outer Lipschitz continuity (originally called upper Lipschitz continuity) was introduced by Robinson [68], where Theorem 3.4 was first stated and proved.

Lecture 4. The name "metric regularity" was coined by J. Borwein in [9], but the origins of this property go back to the Banach open mapping principle. The Aubin property was introduced in Aubin [3] under the name "pseudo-Lipschitz continuity." Theorem 4.4 is from Bessis, Ledyaev and Vinter [8]. For further reading, see Ioffe [48, Section 2].

Lecture 5. Theorem 5.1 stems from two basic results obtained by L. Lyusternik [59] and L. M. Graves [45] with somewhat similar proofs but different motivations. Lyusternik apparently viewed his theorem as a stepping stone to obtain the Lagrange multiplier rule for abstract minimization problems, and the title of his paper clearly says so. It is quite likely that Graves aimed at extending the Banach open mapping theorem for nonlinear mappings. The role of these theorems in variational analysis and optimization was unveiled in the influential paper [17] by Dmitruk, Milyutin and Osmolovskiĭ, for more see Ioffe [48]. Theorem 5.4 is from [27].

Lecture 6. The Robinson–Ursescu Theorem 6.5 was obtained independently by Robinson [66] and Ursescu [77]. It is stated in various ways in the literature; for further reading, see the versions of this theorem in [4] and [73].

Lecture 7. The graphical derivative of set-valued mappings was introduced in Aubin in [2]; for historical comments, see [4]. The proof of Theorem 7.3 is from [41]. Various forms and extensions of the Ekeland variational principle are discussed in detail in the book [10] by Borwein and Zhu. A broad review of the role of coderivatives in variational analysis is given in the books by Rockafellar and Wets [73], Ioffe [48] and Mordukhovich [61, 62], which are recommended for further reading. Theorem 7.6 is from [25], see also [32, Section 4C].

Lecture 8. The property of strong regularity of set-valued mappings was introduced by Robinson [67] in the context of variational inequalities, who also proved in the same paper a version of Theorem 8.3. Various extensions of Robinson's theorem, including prototypes of Theorems 8.5 and Theorem 8.8, were given later in [20]. For further reading, see the books [32] and [48].

Lecture 9. Reduction Lemma goes back to Robinson [69]. Theorem 9.7 was proved in Dontchev and Rockafellar [30], see also Ioffe [48, Chapter 9]. Theorem 9.8 goes back to Kenderov [54].

Lecture 10. Clarke inverse function Theorem 10.1 is from Clarke [13], see also Clarke [14]. The mean value theorem used in its proof is Theorem 2.6.5 in Clarke [13]. The strict graphical derivative was introduced by L. Thibault in [76]. Theorem 10.2 is from Kummer [56]. Theorem 10.4 was originally stated in Izmailov [50] but a complete proof was later given in [11], see also [32, Theorem 4D.4]. For further reading, consider the books by Clarke [14] and Klatte and Kummer [55].

Lecture 11. Theorem 11.1 goes back to [18] while Theorem 11.3 is from [26]. Theorems 11.4 and 11.5 come from Robinson [67]; various versions of these results are available in the literature. The proofs given here are from [32, Sections 2G and 4I]. For further reading, see Klatte and Kummer [55].

Lecture 12. Subregularity, semiregularity, strong subregularity and isolated calmness have been considered in various contexts and under various names in the literature. The name "strong metric subregularity" was first used in [31], where its equivalence with the isolated calmness was proved. Isolated calmness was in turn formally introduced in [19] under the name "local upper Lipschitz continuity at a point"; in the same paper the stability of this property with respect to perturbations was first proved. The equivalent property of strong subregularity was considered earlier, without giving it a name, by Rockafellar [71]. The property of strong subregularity was first considered in [71] who proved there a version of Theorem 12.7. For further reading, see the paper [12].

Lecture 13. Theorem 13.2 is close to the original statement of the this result in Bartle and Graves [5]. Theorem 13.5 is from [22], see also [23], while Theorem 13.7 is from [24].

Lecture 14. The Eckart-Young theorem was published in [42]. Theorems 14.3 and 14.4 are from Dontchev, Lewis and Rockafellar [40]. For more recent results in this direction, see [36] and [38].

Lecture 15. The Newton method applied to variational inequalities was first considered by Josephy [53]; its SQP version remains one of most popular methods for nonlinear programming, see, e.g., the book [51] by Izmailov and Solodov. Theorem 15.1 goes back to [21] while Theorem 15.2 is from [32, Section 6]. The semismooth Newton method for equations was first proposed by Qi and Sun [64], where superlinear convergence of the method was proved. Theorem 15.3 was proved in Izmailov, Kurennoy and Solodov [52]. see also [32, Theorem 6F.1]. For further reading, see the book [55].

Lecture 16. The unconstrained linear-quadratic optimal control problem is broadly covered in most textbooks on optimal control. To a novice in optimal control, the author would recommend to start with the book [60] by Macky and Strauss, and then read the book [78] by Vinter. The double integrator problem is extensively covered in Locatelli [57].

Lecture 17. Theorem 17.1 stems from [33] while Theorem 17.2 follows [26]. More about regularity of mapping associated with optimal control problems that are linear in control, such as in the example at the end of the lecture, see [65]. For further reading, consider [28], where problems involving state constraints are analyzed.

Lecture 18. Theorem 18.2 is a simplified version of a result given in [26]. For higher-order discrete approximations, see [35]. A survey on earlier works about discrete approximations of differential inclusions including optimal control problems is given in [29].

Lecture 19. The main Theorem 19.2 is an adaptation to the linear-quadratic problem of a more general result given in [37].

Lecture 20. The material in this lecture is a simplified presentation of results published in [39]. For further reading, see the book [46] by Grüne and Panek.

References

1. H. Attouch, G. Buttazzo, G. Michaille, *Variational Analysis in Sobolev and BV Spaces. Applications to PDEs and Optimization*, 2nd edn. (SIAM, Philadelphia, 2014)
2. J.P. Aubin, Contingent derivatives of set-valued maps and existence of solutions to nonlinear inclusions and differential inclusions, in *Mathematical Analysis and Applications, Part A*, ed. by L. Nachbin. Advances in Mathematics: Supplementary Studies, 7A (Academic Press, London, 1981), pp. 159–229
3. J.P. Aubin, Lipschitz behavior of solutions to convex optimization problems. Math. Oper. Res. **9**, 87–111 (1984)
4. J.P. Aubin, H. Frankowska, *Set-Valued Analysis* (Birkhäuser, Basel, 1990)
5. R.G. Bartle, L.M. Graves, Mappings between function spaces. Trans. AMS **72**, 400–413 (1952)
6. G. Beer, *Topologies on Closed and Closed Convex Sets* (Kluwer, Dordrecht, 1993)
7. C. Berge, *Topological Spaces, Including a Treatment of Multi-Valued Functions. Vector Spaces and Convexity*. Translated from the French 1959 original (Oliver & Boyd, Edinburgh, 1963)
8. D.N. Bessis, Yu. S. Ledyaev, R.B. Vinter, Dualization of the Euler and Hamiltonian inclusions. Nonlin. Anal. **43**, 861–882 (2001)
9. J.M. Borwein, Stability and regular points of inequality systems. J. Optim. Theory Appl. **48**, 9–52 (1986)
10. J.M. Borwein, Q.J. Zhu, *Techniques of Variational Analysis* (Springer, Berlin, 2005)
11. R. Cibulka, A.L. Dontchev, A nonsmooth Robinson's inverse function theorem in Banach spaces. Math. Program. **156**, 257–270 (2016)
12. R. Cibulka, A.L. Dontchev, V.Y. Kruger, Strong metric subregularity of mappings in variational analysis and optimization. J. Math. Anal. Appl. **457**, 1247–1282 (2018)
13. F.H. Clarke, On the inverse function theorem. Pacific J. Math. **64**, 97–102 (1976)

© The Author(s), under exclusive license to Springer Nature Switzerland AG 2021
A. L. Dontchev, *Lectures on Variational Analysis*, Applied Mathematical Sciences 205, https://doi.org/10.1007/978-3-030-79911-3

14. F.H. Clarke, *Optimization and Nonsmooth Analysis* (John Wiley & Sons, Hoboken, 1983)
15. F.H. Clarke, *Functional Analysis, Calculus of Variations and Optimal Control* (Springer, Berlin, 2013)
16. F.H. Clarke, Yu. S. Ledyaev, R.J. Stern, P.R. Wolenski, *Nonsmooth Analysis and Control Theory* (Springer, Berlin, 1998)
17. A.V. Dmitruk, A.A. Milyutin, N.P. Osmolovskiĭ, Lyusternik theorem and the theory of extremum. Uspekhi Mat. Nauk **35**, 11–46 (1980) (Russian)
18. A.L. Dontchev, Efficient estimates of the solutions of perturbed control problems. J. Optim. Th. Appl. **35**, 85–109 (1981)
19. A.L. Dontchev, Characterizations of Lipschitz stability in optimization, in *Recent Developments in Well-Posed Variational Problems*, ed. by R. Lucchetti, J. Revalski (Kluwer, Dordecht, 1995), pp. 95–115
20. A.L. Dontchev, Implicit function theorems for generalized equations. Math. Program. Ser. A **70**, 91–106 (1995)
21. A.L. Dontchev, Local convergence of the Newton method for generalized equations. C. R. Acad. Sci. Paris **322** Série I, 327–331 (1996)
22. A.L. Dontchev, A local selection theorem for metrically regular mappings. J. Convex Anal. **11**, 81–94 (2004)
23. A.L. Dontchev, Bartle-Graves theorem revisited. Set-Valued Var. Anal. **28**, 109–122 (2020)
24. A.L. Dontchev, On the existence of solutions of parameterized generalized equations. Set-Valued Var. Anal. https://doi.org/10.1007/s11228-020-00554-0
25. A.L. Dontchev, H. Frankowska, On derivative criteria for metric regularity, in *Computational and Analytical Mathematics*, ed. by D.H. Bailey, H.H. Bauschke, P. Borwein, F. Garvan, M. Théra, J. Vanderwerff, H. Wolkowicz, vol. 50 (Springer, Berlin, 2013), pp. 365–374
26. A.L. Dontchev, W.W. Hager, Lipschitzian stability in nonlinear control and optimization. SIAM J. Control Optim. **31**, 569–603 (1993)
27. A.L. Dontchev, W.W. Hager, An inverse mapping theorem for set-valued maps, *Proc. AMS* **121**, 481–489 (1994)
28. A.L. Dontchev, W.W. Hager, Lipschitzian stability for state constrained nonlinear optimal control. SIAM J. Control Optim. **36**, 698–718 (1998)
29. A.L. Dontchev, F. Lempio, Difference methods for differential inclusions – a survey. SIAM Rev. **34**, 263–294 (1992)
30. A.L. Dontchev, R.T. Rockafellar, Characterizations of strong regularity for variational inequalities over polyhedral sets. SIAM J. Optim. **6**, 1087–1105 (1996)
31. A.L. Dontchev, R.T. Rockafellar, Regularity and conditioning of solution mappings in variational analysis. Set-Valued Anal. **12**, 79–109 (2004)
32. A.L. Dontchev, R.T. Rockafellar, *Implicit Functions and Solution Mappings*, 2nd edn. (Springer, Berlin, 2014)
33. A.L. Dontchev, V.M. Veliov, Regularity properties of mappings in optimal control, in *Proceedings of the Conference on Systems Dynamics and Control Processes*, Ekaterinburg (2014), pp. 35–41

34. A.L. Dontchev, T. Zolezzi, *Well-Posed Optimization Problems*. Lecture Notes in Mathematics, vol. 1543 (Springer, Berlin, 1993)
35. A.L. Dontchev, W.W. Hager, V.M. Veliov, Second-order Runge-Kutta approximations in control constrained optimal control. SIAM J. Numer. Anal. **38**, 202–226 (2000)
36. A.L. Dontchev, A. Eberhard, R.T. Rockafellar, Radius theorems for monotone mappings. Set-Valued Var. Anal. **27**, 605–621 (2019)
37. A.L. Dontchev, M.I. Krastanov, V.M. Veliov, On the existence of Lipschitz continuous optimal feedback control. Vietnam J. Math. **47**, 579–597 (2019)
38. A.L. Dontchev, H. Gfrerer, A.Y. Kruger, J. V. Outrata, The radius of metric subregularity. Set-Valued Var. Anal. **28**, 451–473 (2020)
39. A.L. Dontchev, I.V. Kolmanovsky, M.I. Krastanov, V.M. Veliov, P.T. Vuong, Approximating optimal finite horizon feedback by model predictive control. Syst. Control Lett. **139**, 104666 (2020)
40. A.L. Dontchev, A.S. Lewis, R.T. Rockafellar, The radius of metric regularity. Trans. AMS **355**, 493–517 (2003)
41. A.L. Dontchev, M. Quincampoix, N. Zlateva, Aubin criterion for metric regularity. J. Convex Anal. **13**, 281–297 (2006)
42. C. Eckart, G. Young, The approximation of one matrix by another of lower rank. Psychometrica **1**, 211–218 (1936)
43. I. Ekeland, R. Temam, *Analyse convexe et problémes variationnels* (French). Collection Études Mathrńatiques. Dunod; Gauthier-Villars, Paris-Brussels-Montreal, Que., 1974. English translations: Elsevier 1976 (SIAM, Philadelphia, 1999)
44. I.V. Girsanov, *Lectures on Mathematical Theory of Extremum Problems* (Moscow University Press, Moscow, 1970) (Russian). English translation: Lecture Notes in Economics and Mathematics Systems, vol. 67 (Springer, Berlin, 1972)
45. L.M. Graves, Some mapping theorems. Duke Math. J. **17**, 111–114 (1950)
46. L. Grüne, J. Panek, *Nonlinear Model Predictive Control*, 2nd edn. (Springer, Berlin, 2018)
47. A.J. Hoffman, On approximate solutions of systems of linear inequalities. J. Res. Nat. Bureau Stand. **49**, 263–265 (1952)
48. A.D. Ioffe, *Variational Analysis of Regular Mappings* (Springer, Berlin, 2017)
49. A.D. Ioffe, V.M. Tikhomirov, *Theory of Extremal Problems* (Nauka, Moscow, 1974) (Russian). English Translation (North-Holland Publishing Co., Amsterdam, 1979)
50. A.F. Izmailov, Strongly regular nonsmooth generalized equations. Math. Program. A **147**, 581–590 (2014)
51. A.F. Izmailov, A.F. Solodov, *Newton-Type Methods for Optimization and Variational Problems* (Springer, Berlin, 2014)
52. A.F. Izmailov, A.S. Kurennoy, A.F. Solodov, The Josephy-Newton method for semismooth generalized equations and semismooth SQP for optimization. Set-Valued Var. Anal. **21**, 17–45 (2013)

53. N.H. Josephy, Newton's method for generalized equations and the PIES energy model. Ph.D. Dissertation, Department of Industrial Engineering, University of Wisconsin, Madison (1979)
54. P. Kenderov, Semi-continuity of set-valued monotone mappings. Fundamenta Mathematicae **88**, 61–69 (1975)
55. D. Klatte, B. Kummer, *Nonsmooth Equations in Optimization. Regularity, Calculus, Methods and Applications* (Kluwer, Dordecht, 2002)
56. B. Kummer, An implicit-function theorem for $C^{0,1}$-equations and parametric $C^{1,1}$-optimization. J. Math. Anal. Appl. **158**, 35–46 (1991)
57. A. Locatelli, *Optimal Control of a Double Integrator* (Springer, Berlin, 2017)
58. R. Lucchetti, *Convexity and Well-Posed Problems* (Springer, Berlin, 2006)
59. L.A. Lyusternik, On the conditional extrema of functionals. Mat. Sbornik **41**, 390–401 (1934) (Russian)
60. J. Macki, A. Strauss, *Introduction to Optimal Control Theory* (Springer, Berlin, 1982)
61. B.S. Mordukhovich, *Variational Analysis and Generalized Differentiation,* Parts I and II (Springer, Berlin, 2006)
62. B.S. Mordukhovich, *Variational Analysis and Applications* (Springer, Berlin, 2018)
63. J.-P. Penot, *Calculus Without Derivatives* (Springer, Berlin, 2013)
64. L. Qi, J. Sun, A nonsmooth version of Newton's method. Math. Program. A **59**, 353–367 (1993)
65. M. Quincampoix, T. Scarinci, V.M. Veliov, On the metric regularity of affine optimal control systems. J. Convex Anal. **27**, 511–535 (2020)
66. S.M. Robinson, Regularity and stability for convex multivalued functions. Math. Oper. Res. **1**, 130–143 (1976)
67. S.M. Robinson, Strongly regular generalized equations. Math. Oper. Res. **5**, 43–62 (1980)
68. S.M. Robinson, Some continuity properties of polyhedral miltifunctions. Math. Program. Study **14**, 206–214 (1981)
69. S.M. Robinson, An implicit function theorem for a class of nonsmooth functions. Math. Oper. Res. **16**, 292–309 (1991)
70. R.T. Rockafellar, *Convex Analysis* (Princeton University Press, Princeton, 1970)
71. R.T. Rockafellar, Proto-differentiability of set-valued mappings and its applications in optimization. Annales de l'Institut Henri Poincaré, Analyse Non Linéaire **6** suppl., 448–482 (1989)
72. R.T. Rockafellar, Fundamentals of Optimization. Department of Mathematics, University of Washington (2007). Available at http://sites.math.washington.edu/~rtr/mypage.html
73. R.T. Rockafellar, R.J.-B. Wets, *Variational Analysis* (Springer, Berlin, 1998)
74. J.O. Royset, Set-convergence and Its application: a tutorial. Set-Valued Var. Anal. **28**, 707–732 (2020)
75. W. Schirotzek, *Nonsmooth Analysis* (Springer, Berlin, 2007)
76. L. Thibault, Subdifferentials of compactly Lipschitz vector-valued functions. Annali di Matematica Pura Appl. **4**, 157–192 (1980)

77. C. Ursescu, Multifunctions with convex closed graph. Czechoslovak Math. J. **25**, 438–441 (1975)
78. R. Vinter, *Optimal Control* (Birkhäuser, Basel, 2000)
79. D.W. Walkup, R.J.-B. Wets, A Lipschitzian characterization of convex polyhedra. Proc. AMS **23**, 167–173 (1969)
80. C. Zalinescu, *Convex Analysis in General Vector Spaces* (World Scientific, Singapore, 2002)

List of Symbols

$I\!R$: The real numbers
$I\!R_+$: The nonnegative reals
a_+: The positive part of a, vector with components the positive paers of the components of a
$\{x^k\}$: A sequence with elements x^k
$\varepsilon_k \searrow 0$: A sequence of positive numbers ε_k tending to 0
(X, ρ): Metric space with metric $\rho(\cdot, \cdot)$
$|x|$: Absolute value
$||x||$: Any norm
$\langle x, y \rangle$: Canonical inner product, bilinear form
$f : X \to Y$: A function acting from X and Y
$F : X \rightrightarrows Y$: A set-valued mapping acting from X and Y
$\mathcal{L}(X, Y)$: The space of linear and bounded mappings actng between Banach spaces X and Y
$x \perp y$: x is orthogonal to y
$||H||^+$: Outer norm
$||H||^-$: Inner norm
$I\!B_a(x)$: Closed ball with center x and radius r
$I\!B$: Closed unit ball
int $I\!B$: Open unit ball
cl C: Closure
int C: Interior
co$C$: Convex hull
core C: Core
$o(t)$: A quantity satisfying $o(t)/t \to 0$ as $t \searrow 0$
I: Identity mapping, unit matrix
P_C: Projection mapping
$T_C(x)$: Tangent cone
$N_C(x)$: Normal cone
K^*: Polar to cone K, mapping adjoint to K, space dual to K

© The Author(s), under exclusive license to Springer Nature Switzerland AG 2021

A. L. Dontchev, *Lectures on Variational Analysis*, Applied Mathematical Sciences 205, https://doi.org/10.1007/978-3-030-79911-3

$K_C(x, v)$: Critical cone
A^{T}: Transposition
rank A: Rank
ker A: Kernel
det A: Determinant
$d_C(x)$, $d(x, C)$: Distance from x to C
$e(C, D)$: The excess of C beyond D
$h(C, D)$: Pompeiu-Hausforff distance
dom F: Domain
rge F: Range
gph F: Graph
$\nabla f(x)$: Jacobian
$Df(x)$: Derivative
$f'(x, w)$: Directional derivative
$\partial f(x)$: Subdifferential
$\bar{\partial} f(x)$: Clarke generalized Jacobian
$DF(x \,|\, y)$: Graphical derivative
$D^*F(x \,|\, y)$: Coderivative
$D_*F(x \,|\, y)$: Strict graphical derivative
$\mathrm{clm}\,(f; x)$, $\mathrm{clm}\,(S; y \,|\, x)$: Calmness modulus
$\mathrm{lip}\,(S; y \,|\, x)$: Lipschitz modulus of a set-valued mapping
$\mathrm{lip}\,(f; x)$: Lipschitz modulus of a function
$\widehat{\mathrm{clm}}_p(f; (p, x))$: Partial calmness modulus
$\widehat{\mathrm{lip}}_p(f; (p, x))$: Partial Lipschitz modulus
$\mathrm{reg}\,(F; x \,|\, y)$: Regularity modulus
$\mathrm{subreg}\,(F; x \,|\, y)$: Subregularity modulus

Index

© The Author(s), under exclusive license to Springer Nature Switzerland AG 2021

A. L. Dontchev, *Lectures on Variational Analysis*, Applied Mathematical Sciences 205, https://doi.org/10.1007/978-3-030-79911-3

GPSR Compliance
The European Union's (EU) General Product Safety Regulation (GPSR) is a set of rules that requires consumer products to be safe and our obligations to ensure this.

If you have any concerns about our products, you can contact us on

ProductSafety@springernature.com

In case Publisher is established outside the EU, the EU authorized representative is:

Springer Nature Customer Service Center GmbH
Europaplatz 3
69115 Heidelberg, Germany

www.ingramcontent.com/pod-product-compliance
Ingram Content Group UK Ltd.
Pitfield, Milton Keynes, MK11 3LW, UK
UKHW021947270726
14060UKWH00002B/406

* 9 7 8 3 0 3 0 7 9 9 1 0 6 *